W9-BNX-014

Current Topics in Microbiology and Immunology

266

Springer
Berlin
Heidelberg
New York
Barcelona
Hong Kong
London
Milan
Paris
Tokyo

The Interface Between Innate and Acquired Immunity

Edited by M.D. Cooper
and H. Koprowski

With 15 Figures and 3 Tables

Springer

Professor Dr. MAX D. COOPER
University of Alabama at Birmingham
Howard Hughes Medical Institute
Research Laboratories
378 Wallace Tumor Institute
Birmingham, Alabama 35294, USA
E-mail: max.cooper@ccc.uab.edu

Professor Dr. HILARY KOPROWSKI
Jefferson Medical College
Department of Microbiology and Immunology
Center for Neurobiology, Jefferson Alumni Hall
1020 Locust Street, Philadelphia, PA 19197-6799, USA
E-mail: hkoprowski@reddil.uns.tju.edu

Cover Illustration: The phylogenetic relationship between the recently discovered Fc receptor homologs and classical Fc receptor family members is indicated by the color-coded depiction of their extracellular Ig-like domains and a corresponding phylogenetic tree (Fig. 4 of Davis et al., this volume). Each domain is color-coded in a manner corresponding to its clustering position. The ClustalX program was used for comparative analysis of an all-against-all amino acid sequence alignment of the Ig-like domains. The unrooted tree topology was estimated using TreeCon. The branch values, representing percentage Bootstrap support after 500 replicates, indicate a high degree of confidence in the validity of the relationships.

ISSN 0070-217X
ISBN 3-540-42894-1 Springer-Verlag Berlin Heidelberg New York

Springer-Verlag Berlin Heidelberg New York
a member of BertelsmannSpringer Science + Business Media GmbH

http://www.springer.de

Library of Congress Catalog Card Number 15-12910
Printed in Germany

Cover Design: *design & production GmbH*, Heidelberg
Typesetting: Scientific Publishing Services (P) Ltd, Madras
Printed on acid-free paper SPIN: 10718419 27/3020 - 5 4 3 2 1 0

Preface

All multicellular organisms may possess innate immunity mediated by defense mechanisms with which the organism is born. In recent years much has been learned about the diversity of innate immune mechanisms. A large array of naturally produced antimicrobial peptides has been defined. A variety of cell surface receptors that recognize common patterns displayed by infectious organisms have been identified along with the intracellular pathways that these receptors use to activate cellular defense functions. Cell surface receptors on natural killer (NK) cells have been shown to sense microbial invasion in neighboring cells, thereby setting into motion their elimination by cytotoxic mechanisms. Other receptors have been found to facilitate phagocytosis and intracellular killing of microbes by phagocytic cells. These and other natural defense mechanisms have traditionally been viewed as the first line of body defense in vertebrate species that also possess the capacity for acquired or adaptive immunity. Sharks and all of the other jawed vertebrates generate large repertoires of T and B lymphocyte clones that display different antigen-specific receptors in the form of T cell receptors (TCR) and immunoglobulins (Ig) that allow them to recognize and respond to antigens in collaboration with antigen-presenting cells. Memory T and B cells are then generated to allow faster and heightened cellular and humoral immune responses on secondary antigen encounter.

In recent years it has also become obvious that innate immune responses can directly influence adaptive immune responses in ways that will enhance body defense. In turn, the cytokine and antibody products of responding T and B cells can invoke or heighten the participation of inflammatory and phagocytic cell members of the innate immune system. The complexity of the multifaceted interfaces between innate and adaptive immunity was the meeting topic of a group of European and North American investigators last year in Malta, not far from where Metchnikoff first recognized phagocytosis by inflammatory cells in starfish larvae responding to a thorn injury. The reports in this volume represent contributions by some of the Malta conference participants. The topics include: the regulation of the helper T cell subset directly involved in inflammatory responses (Th1 cells) versus the subset of helper T cells involved in promoting B cell responses in humoral immunity (Th2 cells); the regulation of T cell activity in inflammatory and autoimmune disease by the tumor necrosis factor (TNF) superfamily of receptors and their ligands; the extensive array of inhibitory and activating receptors that regulate NK cell function; the phylogeny of the immunoglobulin gene superfamily members linked to the major histocompatibility (MHC) gene complex; the immunoglobulin Fc receptor family and a newly discovered family of Fc receptor homologs; the role of complement in linking innate and adaptive immunity; and manipulation of the potential for plants to protect themselves by innate and adaptive responses to plant viruses.

Anne O'Garra and her colleagues address mechanisms that are involved in driving helper T cell differentiation along the Th1 and Th2 pathways. Interleukin (IL)-12 is the key cytokine that drives Th1 cell development, but this process may be augmented by IL-1α and IL-18 as evidence for the complexity of this process. These authors elaborate on the role of signal transduction and activation of transcription factor 6 (STAT6) in enhancing Th1 cell activity and the role of the GATA-3 transcription factor in determining Th2 T cell differentiation.

Richard Flavell discusses the role that members of the TNF receptor/ligand families play in the breakdown of tolerance and activation of autoreactive T cells that results in an autoimmune attack on host tissues. This contribution provides support for the hypothesis that factors which influence the degree of host cell apoptosis can be a major determinant in whether or not autoimmunity occurs. The apoptosis of insulin-producing β islet cells associated with an inflammatory response in the pancreas can thus lead to cross priming of T cells and their participation in the development of diabetes. Flavell presents a model in which immature dendritic cells in the presence of an inflammatory response are drawn into the pancreas, wherein the dendritic cells pick up β-cell peptides and present these to autoreactive T cells in a way that breaks down tolerance. The activated T cells then migrate into the pancreatic islets where they destroy the β-cells.

NK cells can immediately recognize virus-infected cells and lyse them, but at the same time they must be prevented from killing normal neighboring cells. Although the NK cells derive from common lymphoid progenitors of the T and B cells, NK cells lack the specific antigen receptors that are created by the V(D)J gene segment rearrangements in T and B cell precursors. In fact, the NK cells are thought to represent components of the innate immune system that arose in invertebrates. For all of these reasons there has been an intense interest in determining the nature of the receptors that NK cells use to discriminate between normal healthy cells and host cells invaded by pathogenic microbes. Lorenzo and Alessandro Moretta, leaders in the field of NK receptor biology, describe the relatively large array of activating and inhibitory receptors that have now been identified on NK cells. Many of these NK cell receptors, but not all, recognize MHC class I ligands.

The paired expression of a diverse group of activating and inhibitory Ig-like receptors can serve to regulate the activation status of NK cells, T cells, B cells and other types of cells involved in immune and inflammatory responses. A large complex of Ig-gene superfamily members encoding these types of receptors are located on the long arm of chromosome 19 in humans and in the syntenic region of chromosome 7 in mice. Randall Davis, Glynn Dennis, and their colleagues have conducted a phylogenetic analysis that indicates a relationship between this paired Ig-like receptor family and the classical Ig Fc receptors that are encoded on the long arm of human chromosome 1. A search for other gene relatives in this region led to the identification of a family of five receptors that possess variable numbers of extracellular Ig domains and either activating motifs or inhibitory motifs or both in their cytoplasmic domains. These investigators have provisionally named members of the newly identified family the Fc receptor homologs (FcRH1-5) because of shared extracellular Ig domain homology with classical FcR family members, sharing of specific sequence homology within the Fc binding site of the FcRs, and location of the *FcRH* genes in the midst of the *FcR* genes on chromosome 1q21. This receptor family has also been identified through an analysis of a chromosomal translocation breakpoint in a multiple myeloma cell line. The two different strategies used to identify these genes provide clues to their likely biological roles as FcR related molecules with diverse signaling characteristics and oncogenic potential. Their extracellular features, signaling potentials, and preferential B cell expression patterns suggest the Fc receptor homologs are differentially involved in regulating B cell responses. The potential for FcR and FcRHs to link natural antibodies to effector cells could provide yet another interface between innate and adaptive immunity.

Identification of the phylogenetic ancestors of rearranging Ig and TCR genes is the focus of Louis du Pasquier, an authority on immune system phylogeny. So far this phylogenetic analysis of genes that encode variable (V) region, constant region type I (C1), and combined V-C1 domains suggests that the C1 Ig domain structure is restricted to jawed vertebrates that employ somatic V(D)J rearrangements for generating their antigen specific receptors. The analysis of V-C1 and single V genes localizes two linkage groups including the MHC complex of genes that may have served as the source of an ancestral receptor gene.

The complement system of proenzymes constitutes a relatively ancient type of innate immunity in that it is present in all vertebrates and complement components have been identified in some invertebrate species. Complement components can recognize conserved repetitive structural elements on pathogens, and the resulting activation of a cascade of complement proenzymes may destroy the pathogens by lysis, via the attraction of inflammatory cells, or through coating the pathogens before binding to complement receptors on phagocytic cells that then ingest and digest the pathogen. The many different ways of activating complement and the role of this ancient system in bridging innate and adaptive immunity are cogently reviewed by John Volanakis.

Plants may also use innate and acquired immune strategies for repelling microbial invaders. Pathogen-derived resistance (PDR) is an example of acquired immunity in that plants which express a viral gene may become resistant to that particular virus. Dennis Gonzales examines the practical application of PDR to control a deadly viral infection in papaya plants. Posttranscriptional gene silencing is shown to be the underlying mechanism of the RNA-mediated resistance that can be conferred by the production of transgenic papaya plants, and he discusses how this acquired immunity may be used successfully to control papaya ring spotted viral infections in papaya. This strategy of disease prevention could have general applicability as a means for protection against other plant pathogens.

Hilary Koprowski and Max Cooper

List of Contents

List of Contributors

(Their addresses can be found at the beginning of their respective chapters.)

The Relationship of Inflammation and Initiation of Autoimmune Disease: Role of TNF Super Family Members

R.A. FLAVELL

1 Introduction

In normal people, autoreactive T cells can be deleted, either in the thymus or in the periphery, following presentation of self-peptides by host antigen-presenting cells (APCs) to self-reactive T cells (KURTS et al. 1996, 1998a,b). Breakdown in this putative tolerance mechanism is believed to contribute to autoimmunity in genetically susceptible individuals. Type I diabetes mellitus is characterized by infiltration of the islets of Langerhans by immune cells, which ultimately destroy the insulin-producing β-cells by T cell-mediated mechanisms. The T cell-mediated nature of this autoimmunity requires that APCs not only have to present islet antigen released from β-cells but also have to deliver signals that promote survival of the self-reactive T cells. Identification of the cells and events that initiate and maintain this anti-islet inflammatory response is vital in the development of therapeutic strategies towards diabetes.

Section of Immunobiology, Yale University School of Medicine, and Howard Hughes Medical Institute, New Haven, CT 06520-8011, USA

2 Apoptosis and the Initiation of Insulitis

The islet antigens to which the initial T cell responses are directed have not been identified, although glutamic acid decarboxylase or insulin are strong candidates (Wong et al. 1999; Yoon et al. 1999). The availability of islet antigen for T cell priming is likely to derive from immune attack on β-cells. In particular, viral infections that cause pancreatic inflammation have been implicated as perpetrators of this initial β-cell damage in children (Andreoletti et al. 1998; Horwitz et al. 1998).

In NOD mice, it is unlikely that viral infection contributes to the provision of islet antigen for T cell priming, since diabetes susceptibility requires a specific pathogen-free environment (Delovitch and Singh 1997). We have proposed that an alternative explanation of how islet antigens are made available to the immune system is apoptosis (O'Brien et al. 1997; Green et al. 1998). Apoptosis is the process whereby old or defective cells are removed from the body. In most instances, apoptotic cells are rapidly cleared from tissues by scavenger macrophages to prevent inappropriate inflammatory responses. However, if apoptosis occurs in the presence of a strong inflammatory response, these immature dendritic cells (DCs) receive maturational signals enabling them to cross-prime T cells (Albert et al. 1998). We provided support for the hypothesis of apoptosis as the mechanism by which diabetes may be initiated. We showed that in tumor necrosis factor (TNF)-α-NOD mice (which express TNF-α in their islets from birth) β-cell apoptosis precedes infiltration of the islets with immature DCs, and in this inflammatory environment the DCs became activated (Green et al. 1998).

Although it is likely that apoptosis of β-cells in association with an inflammatory response in the islets can lead to cross-priming of T cells, how does this relate to the development of diabetes in a genetically unmanipulated NOD mouse? Apoptosis of β-cells occurs naturally during remodeling of the pancreas (Scaglia et al. 1997), but it is not known what induces the influx of immune cells that constitute the intra-islet inflammatory response (Voorbij et al. 1989; Jansen et al. 1993; Rosmalen et al. 1997; Dahlén et al. 1998). The studies of Rovere et al. (1998, 1999) may hold the key to this dilemma. In an in vitro model, they showed that there was a threshold for the number of apoptotic cells that could be cleared by macrophages without inducing inflammation. Above this threshold, macrophages became overwhelmed and failed to clear apoptotic cells. This caused persistence of apoptotic cells, which were subsequently ingested by immature DCs. The uptake of apoptotic cells by the immature DCs promoted DC activation and the release of inflammatory cytokines like lymphotoxin (LT)-β and TNF-α (Rovere et al. 1998, 1999). This finding may explain why the ability to generate strong anti-tumor cytotoxic T lymphocyte (CTL) responses following immunization with apoptotic tumor cells is proportional to the number of apoptotic cells used for immunization (Ronchetti et al. 1999).

Defects in the clearance of apoptotic cells have been speculated to contribute to autoimmunity in systemic lupus erythematous (SLE) patients (Carroll 1998) and in murine models of SLE (Mevorach et al. 1998). Several reports have documented that lymphocytes in the NOD mouse and BB rat show abnormal resistance or susceptibility to apoptosis (Penha-Gonçalves et al. 1995; Casteels et al. 1998; Hernandez-Hoyos et al. 1999; Martins and Aguas 1999). It will be interesting to see if apoptosis-related mechanisms in the islet contribute to the development of diabetes in the NOD mouse, and potentially in humans. A potential model for the activation of islet-reactive T cells is shown in Fig. 1.

3 APCs and the Initiation of Insulitis

Several investigations have focused on identifying which APCs are responsible for priming islet-specific T cells. Both B cells and DCs have been implicated in the initiation of anti-islet T cell responses. NOD mice treated with anti-Igμ heavy chain-specific antibodies from birth or NOD mice deficient in B cells (NOD-Igμ^{null} mice) fail to develop insulitis (Serreze et al. 1996, 1998; Akashi et al. 1997; Noorchashm et al. 1997) and are protected from diabetes. Protection from diabetes in genetically manipulated NOD mice has also been linked to defects in B cell-antigen presentation (King et al. 1998). On the other hand, we have demonstrated in TNFα-NOD mice an accelerated model of diabetes that DCs present in the earliest islet infiltrates have pre-formed peptide-MHC complexes on their surface and stimulate islet-specific T cells in vitro and potentially in vivo (Green et al. 1998). Diabetes development in TNFα-NOD mice occurs in the absence of B cells, suggesting that DCs are the principal APCs at both the initiation and effector stages of diabetes in this model (E.A. Green and R.A. Flavell, unpublished data). These findings favor DCs as the initiators of the islet-specific T cell response. Interestingly, many reports demonstrate that scavenger macrophages and DCs are the first cells to infiltrate the islets of manipulated and non-manipulated NOD mice (Rosmalen et al. 1997; Dahlén et al. 1998; Voorbij et al. 1989)

Studies comparing the potency of B cells compared with DCs in triggering antigen-dependent Ca^{2+} responses in naïve T cells have shown that

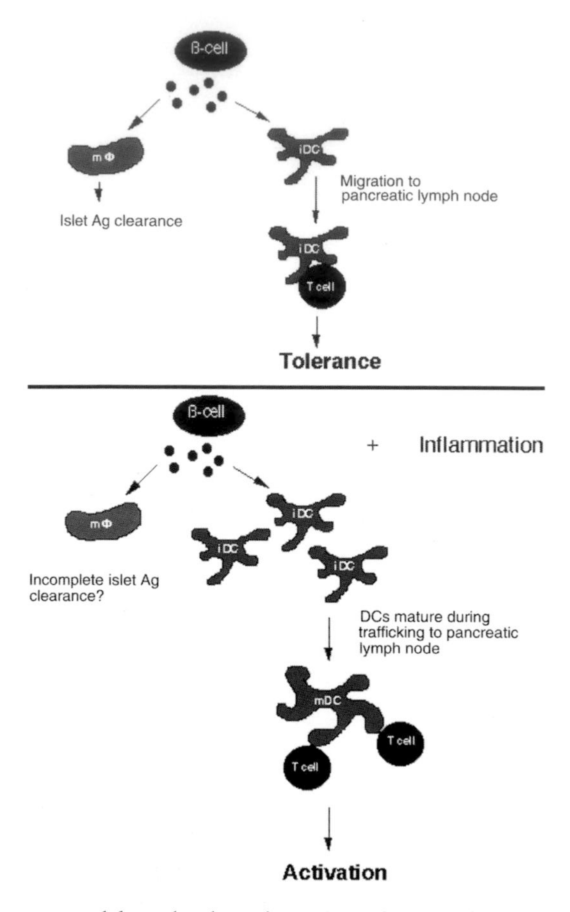

Fig. 1. A model for the activation of islet-reactive T cells. *Upper panel*: In the absence inflammation, both scavenger macrophages (*mΦ*) and immature dendritic cells (*iDC*) acquire β-cell antigens. The scavenger mΦs rapidly clear islet Ag, whereas the iDC migrate to the pancreatic lymph node (*PLN*). Here, the iDCs present islet peptides to autoreactive T cells and induce T cell tolerance. *Lower panel*: In the presence of an inflammatory response, more iDCs are recruited to the islets than scavenger mΦs. These iDCs rapidly acquire islet fragments, and traffic to the PLN. During this migration, the iDCs mature due to the ongoing inflammatory response. Interaction of these mature DCs (*mDCs*) with autoreactive T cells results in T cell activation, not tolerization. Activated T cells then migrate to the islet and, following secondary exposure to islet antigen in situ, destroy β cells

several hundred antigen-MHC complexes on each B cell were required for efficient activation of T cells, whereas only 30 antigen-MHC complexes on each DC promoted T cell responses (Delon et al. 1998). In vivo, the interaction of DCs (bearing antigen) with naïve T cells has been visualized in the draining lymph nodes prior to initiation of antigen-specific T cell responses (Ingulli et al. 1997). Furthermore, immunization with DCs bearing islet antigen can promote diabetes (Ludewig et al. 1998). These data suggest DC as the likely APC in diabetes.

4 Signal Pathways and Autoimmunity

The breakdown in tolerance of islet-specific T cells to their cognate antigen is the defining point in the initiation of diabetes. It is important to understand what signal pathways between APCs and T cells determine whether T cells are tolerized or primed to self-antigens. The original concept that T cell activation requires two signals from the APC, the first generated via interaction of the T cell antigen receptor (TCR) with peptide-MHC complexes and the second through interaction of CD28 or CTLA4 on T cells and B7 co-stimulatory molecules on APCs, is now known to be over-simplistic. Recent advances have demonstrated other important signals between T cells and APCs that govern whether T cells become activated or are tolerized. Some of these signals involve interaction between members of the TNF and TNF receptor (TNFR) families.

The TNF family includes Fas ligand, CD27, CD30, CD154, 4–1BB, OX-40, TNF-α, LT-a/β, TNF-related activation-induced cytokine (TRANCE), and TNF-related activation-induced ligand (TRAIL); there is a high degree of homology between the family members (Kwon et al. 1999). Members of the TNFR family that includes Fas are CD27R, CD30R, CD40, 4–1BBR, OX-40L (OX-40 ligand), TNFRI/II, LT-α/βR, receptor activator of NFκB (RANK), and DR4; similarly, these have strong amino acid identity. Of these family members, two molecules have received considerable attention for their importance in regulating immune responses: CD154, which is expressed on activated $CD4^+$ T cells, and its receptor (CD40), which is expressed on APCs. The importance of CD154-CD40 signals for efficient priming of T cells is exemplified by studies in mice treated with neutralizing CD154 antibodies or deficient in CD154 ($CD154^{-/-}$ mice), where abrogation of the CD154 signal causes defective cell-mediated and humoral immunity to a variety of antigens (Foy et al. 1993; Xu et al. 1994; Grewal et al. 1995; Stout et al. 1996). In humans, mutations in the CD154 gene result in autoimmune hypergammaglobulinemia (Grewal and Flavell 1998). In murine models of experimental autoimmune encephalitis (EAE), multiple sclerosis, and diabetes, the blocking of CD154-CD40 signals can prevent autoimmunity (Grewal et al. 1996; Balasa et al. 1997; Mach et al. 1998).

One important role of the CD154-CD40 interaction is to induce activation of APCs. This was shown independently by three groups, each of which demonstrated that treatment of APCs in vitro or in vivo with anti-CD40 antibodies was sufficient to induce APC activation and the presentation of peptides to T cells (Bennet et al. 1998; Ridge et al. 1998; Schoenberger et al. 1998). In these studies, $CD4^+$ T cell help for $CD8^+$ T cell activation could be bypassed by treatment of APCs with anti-CD40. Thus, $CD4^+$ T cells promote activation of $CD8^+$ T cells by providing the CD154-CD40 signal for APC activation.

Manipulation of CD154-CD40 signals has been considered as a new form of therapy in diabetes. For example, long-term survival of syngeneic islet grafts occurs if recipients are treated with donor splenocytes coupled to anti-CD154 therapy (Gordon et al. 1998). Although anti-CD154 therapy is an attractive idea for the treatment of diabetes, its success depends on the assumption that activation of T cells is solely dependent on CD154-mediated signals. However, under inflammatory conditions APCs can become activated in the absence of CD154-CD40 signals, and these APCs can efficiently prime $CD8^+$ T cells to their cognate antigen. We first demonstrated this for viral infections that induce antiviral CTL responses in the absence of CD154-mediated help from $CD4^+$ T cells (Borrow et al. 1998). In more recent studies a pro-inflammatory environment in the islets of NOD-$CD154^{-/-}$ mice has been shown to promote APC maturation, presentation of islet peptides to $CD8^+$ T cells, and the subsequent development of diabetes (Green et al. 2000). This latter result emphasizes the importance of the immune status of the islet environment when APCs encounter T cells in situ.

The CD154-independent pathways by which APCs can become activated were until recently unknown, although viral infections can promote APC activation in CD40-deficient mice following interaction of TRANCE on T cells and RANK on APCs (Bachmann et al. 1999). TRANCE is structurally similar to CD154; expressed on activated T cells, it can induce activation of DCs and the release of pro-inflammatory cytokines by signaling through its receptor (RANK) on the DC surface (Green and Flavell 1999). The TRANCE–RANK signal is therefore an attractive candidate for the pathway by which inflammatory conditions can cause T cell activation in the absence of CD154-CD40 signals. It will be important to establish the role of TRANCE–RANK signals in autoimmunity.

Members of the TNF and TNFR families, therefore, have distinct roles to play in the regulation of T cell immune responses. With the exception of TNF-α, TNFRs, and CD154, it remains to be seen how important these diverse molecules are in the initiation and effector stages of diabetes in NOD mice.

5 Conclusions

Our understanding of the cellular and molecular events involved in the initiation of diabetes is increasing. Future studies that identify precisely the APC responsible for priming T cells to islet antigen and the signal pathways that diabetogenic T cells utilize to avoid tolerization should aid the design of newer therapeutic strategies.

Acknowledgements. I would like to thank Fran Manzo for help in manuscript preparation. RA Flavell is an Investigator of the Howard Hughes Institute.

Work from this laboratory described here was supported by grants from the National Institutes of Health, Juvenile Diabetes Foundation International and the American Diabetes Association.

References

Akashi T, Nagafuchi S, Anzai K, Kondo S, Kitamura D, Wakana S, Ono J, Kikuchi M, Niho Y, Watanabe T (1997) Direct evidence for the contribution of B cells to the progression of insulitis and development of diabetes in non-obese diabetic mice. Intl Immunol 9:1159–1164

Albert ML, Sauter B, Bhardwaj N (1998) Dendritic cells acquire antigen from apoptotic cells and induce class I-restricted CTLs. Nature 392:86–89

Andreoletti L, Hober D, Nober-Vandenberghe C, Fajardy I, Belaich S, Lambert V, Vantyghem M, Lefebvre J, Wattre P (1998) Coxsackie B virus infection and beta cell autoantibodies in newly diagnosed IDDM adult patients. Clin Diagn Virol 9:125–133

Bachmann MF, Wong BR, Josien R, Steinman RM, Oxenius A, Choi Y (1999) TRANCE (tumor necrosis factor [TNF]-related activation-induced cytokine), a TNF family member critical for CD40 ligand-independent T helper cell activation. J Exp Med 189:1025–1031

Balasa B, Krahl T, Patstone G, Lee J, Tisch R, McDevitt HO, Sarvetnick N (1997) CD40 ligand-CD40 interactions are necessary for the initiation of insulitis and diabetes in nonobese diabetic mice. J Immunol 159:4620–4627

Bennet SRM, Carbone FR, Karamalis F, Flavell RA, Miller JFAP, Heath WR (1998) Help for cytotoxic-T cell responses is mediated by CD40 signaling. Nature 393:478–480

Borrow P, Tough DF, Eto D, Tishon A, Grewal IS, Sprent J, Flavell RA, Oldstone BA (1998) CD40 ligand-mediated interactions are involved in the generation of memory CD8(+) cytotoxic T lymphocytes (CTL) but are not required for the maintenance of CTL memory following virus infection. J Virol 72:7440–7449

Carroll M (1998) The lupus paradox. Nat Genet 19:3–4

Casteels K, Gysemans C, Waer M, Bouillon R, Laureys J, Depovere J, Mathieu C (1998) Sex difference in resistance to dexamethasone-induced apoptosis in NOD mice. Diabetes 47:1033–1037

Dahlén E, Dawe K, Ohlsson L, Hedlund G (1998) Dendritic cells and macrophages are the first and major producers of TNFα in pancreatic islets in the nonobese diabetic mouse. J Immunol 160:3585–3593

Delon J, Bercovici N, Raposo G, Liblau R, Trautmann A (1998) Antigen-dependent and independent Ca2+ responses triggered in T cells by dendritic cells compared with B cells. J Exp Med 188:1473–1484

Delovitch TL, Singh B (1997) The nonobese diabetic mouse as a model of autoimmune diabetes: immune dysregulation gets the NOD. Immunity 7:727–738

Foy TM, Shepherd DM, Durie FH, Aruffo A, Ledbetter JA, Noelle RJ (1993) In vivo CD40-gp39 interactions are essential for thymus-dependent humoral immunity. II. Prolonged suppression of the humoral immune response by an antibody to the ligand for CD40, gp39. J Exp Med 178:1567–1575

Gordon E, Markees T, Phillips N, Noelle R, Shultz L, Mordes J, Rossini A, Greiner D (1998) Prolonged survival of rat islet and skin xenografts in mice treated with donor splenocytes and anti-CD154 monoclonal antibody. Diabetes 47:1199–1206
Green EA, Flavell RA (1999) TRANCE-RANK, a new signal pathway involved in lymphocyte development and T cell activation. J Exp Med 189:1017–1020
Green EA, Eynon EE, Flavell RA (1998) Local expression of TNFalpha in neonatal NOD mice promotes diabetes by enhancing presentation of islet antigens. Immunity 9:733–743
Green E, Wong F, Eshima K, Mora C, Flavell R (2000) Neonatal tumor necrosis factor α promotes diabetes in nonobese diabetic mice by CD154-independent antigen presentation to CD8+ T cells. J Exp Med 191:225–237
Grewal IS, Flavell RA (1998) CD40 and CD154 in cell-mediated immunity. Ann Rev Immunol 16:111–135
Grewal IS, Xu J, Flavell RA (1995) Impairment of antigen-specific T-cell priming in mice lacking CD40 ligand. Nature 378:617–620
Grewal IS, Foellmer HG, Grewal KD, Xu J, Hardardottir F, Baron JL, Janeway CAJ, Flavell RA (1996) Requirement for CD40 ligand in co-stimulation induction, and experimental allergic encephalomyelitis. Science 273:1864–1867
Hernandez-Hoyos G, Joseph S, Miller N, Butcher G (1999) The lymphopenia mutation of the BB rat causes inappropriate apoptosis of mature thymocytes. Eur J Immunol 29:1832–1841
Horwitz MS, Bradley LM, Habertson J, Krahl T, Lee J, Sarvetnick N (1998) Diabetes induced by coxsackie virus: initiation by bystander damage and not molecular mimicry. Nature Medicine 4:781–785
Ingulli E, Mondino A, Khoruts A, Jenkins MK (1997) In vivo detection of dendritic cell antigen presentation to CD4+ T cells. J Exp Med 185:2133–2141
Jansen A, Voorbij HAM, Jeucken PHM, Bruining GJ, Hooijkaas H, Drexhage HA (1993) An immunohistochemical study on organized lymphoid cell infiltrates in fetal and neonatal pancreas. A comparison with similar infiltrates found in the pancreas of a diabetic infant. Autoimmunity 15:31–38
King C, Davies J, Mueller R, Lee M-S, Krahl T, Yeung B, O'Connor E, Sarvetnick N (1998) TGF-β1 alters APC preference, polarizing islet antigen responses toward a Th2 phenotype. Immunity 8:601–613
Kurts C, Heath W, Kosaka H, Miller J, Carbone F (1998a) The peripheral deletion of autoreactive CD8+ T cells induced by cross-presentation of self-antigens involves signaling through CD95 (Fas, Apo-1). J Exp Med 188:415–420
Kurts C, Miller JFAP, Subramanian RM, Carbone FR, Heath WR (1998b) Major histocompatibility complex class I-restricted cross-presentation is biased towards high dose antigen and those released during cellular destruction. J Exp Med 188:409–414
Kurts C, Heath WR, Carbone FR, Allison J, Miller JF, Kosaka H (1996) Constitutive class I-restricted exogenous presentation of self antigens in vivo. J Exp Med 184:923–930
Kwon B, Youn B-S, Kwon B (1999) Functions of newly identified members of the tumor necrosis factor receptor/ligand superfamilies in lymphocytes. Curr Opin Immunol 11: 340–345
Ludewig B, Odermatt B, Landmann S, Hengartner H, Zinkernagel RM (1998) Dendritic cells induce autoimmune diabetes and maintain disease via de novo formation of local lymphoid tissue. J Exp Med 188:1493–1501
Mach F, Shonbeck U, Sukhova GK, Atkinson E, Libby P (1998) Reduction of atherosclerosis in mice by inhibition of CD40 signaling. Nature 394:200–203
Martins T, Aguas A (1999) NOD mice are resistant to depletion of thymic cells caused by acute stress or infection. Autoimmunity 29:273–280
Mevorach D, Zhou J, Song X, Elkon K (1998) Systemic exposure to irradiated apoptotic cells induces autoantibody production. J Exp Med 188:387–392

Noorchashm H, Noorchashm N, Kern J, Rostami SY, Barker CF, Naji A (1997) B-cells are required for the initiation of insulitis and sialitis in nonobese diabetic mice. Diabetes 46:941–946

O'Brien BA, Harmon BV, Cameron DP, Allan DJ (1997) Apoptosis is the mode of β-cell death responsible for the development of IDDM in the nonobese diabetic (NOD) mouse. Diabetes 46:750–757

Penha-Gonçalves C, Leijon K, Persson L, Holmgren D (1995) Type 1 diabetes and the control of dexamethasone-induced apoptosis in mice maps to the same region on chromosome 6. Genomics 28:398–404

Ridge F, Di Rosa F, Matzinger P (1998) A conditioned dendritic cell can be a temporal bridge between a CD4+ T-helper cell and a T-killer cell. Nature 393:474–478

Ronchetti A, Iezzi G, Crosti M, Garancini M, Protti M, Bellone M (1999) Role of antigen-presenting cells in cross-priming of cytotoxic T lymphocytes by apoptotic cells. J Leukoc Biol 66:247–251

Rosmalen JGM, Leenen PJM, Katz JD, Voerman JSA, Drexhage HA (1997) Dendritic cells in the autoimmune insulitis in NOD mouse models of diabetes. Adv Exp Med Biol 417:291–294

Rovere P, Vallinoto C, Bondanza A, Crosti MC, Rescigno M, Ricciardi-Castagnoli P, Rugarli C, Manfredi AA (1998) Cutting edge: bystander apoptosis triggers dendritic cell maturation and antigen-presenting function. J Immunol 161:4467–4471

Rovere P, Sabbadini M, Vallinoto C, Fascio U, Zimmermann V, Bondanza A, Ricciardi-Castagnoli P, Manfredi A (1999) Delayed clearance of apoptotic lymphoma cells allows cross-presentation of intracellular antigens by mature dendritic cells. J Leukoc Biol 66:345–349

Scaglia L, Cahill CJ, Finewood DT, Bonner-Weir S (1997) Apoptosis participates in the remodeling of the endocrine pancreas in the neonatal rat. Endocrinol 138:1736–1741

Schoenberger SP, Toes REM, Voort-van-der EIH, Offringa R, Melief CJM (1998) T-cell help for cytotoxic T lymphocytes is mediated by CD40-CD40L interactions. Nature 393: 480–483

Serreze DV, Chapman HD, Varnum DS, Hanson MS, Reifsnyder PC, Richard SD, Fleming SA, Leiter EH, Shultz LD (1996) B lymphocytes are essential for the initiation of T cell-mediated autoimmune diabetes: analysis of a new "speed congenic" stock of NOD. Igu mu null mice. J Exp Med 184:2049–2053

Serreze DV, Fleming SA, Chapman HD, Richard SD, Leiter EH, Tisch RM (1998) B lymphocytes are critical antigen-presenting cells for the initiation of T cell-mediated autoimmune diabetes in nonobese diabetic mice. J Immunol 161:3912–3918

Stout RD, Suttles J, Xu J, Grewal IS, Flavell RA (1996) Impaired T cell-mediated macrophage activation in CD40 ligand-deficient mice. J Immunol 156:8–11

Voorbij HAM, Jeucken PHM, Kabel PJ, De Haan M, Drexhage HA (1989) Dendritic cells and scavenger macrophages in pancreatic islets of prediabetic BB rats. Diabetes 38:1623–1629

Wong F, Karttunen J, Dumont C, Wen L, Visintin I, Pilip I, Shastri N, Pamer E, Janeway CJ (1999) Identification of an MHC class I-restricted autoantigen in type I diabetes by screening an organ-specific cDNA library. Nat Med 5:1026–1031

Xu J, Foy TM, Laman JD, Elliott EA, Dunn JJ, Waldschmidt TJ, Elsemore J, Noelle RJ, Flavell RA (1994) Mice deficient for the CD40 ligand. Immunity 1:423–431

Yoon J, Yoon C, Lim H, Huang Q, Kang Y, Pyun K, Hirasawa K, Sherwin R, Jun H (1999) Control of autoimmune diabetes in NOD mice by GAD expression or suppression in beta cells. Science 284:1183–1187

Surface Receptors that Regulate the NK Cell Function: Beyond the NK Cell Scope

L. Moretta[1,2], R. Biassoni[3], C. Bottino[3], M.C. Mingari[3,4], and A. Moretta[2]

1 Introduction

Natural killer (NK) cells were originally described on a functional basis, according to their ability to lyse certain tumors in the absence of prior stimulation. However, neither their origin nor the molecular mechanism(s) involved in their function had been defined (Trinchieri 1989). Regarding the origin of the NK cells, clear evidence has been provided both in mouse and in man that NK and T cells may derive from a common precursor (Mingari et al. 1991; Rodewald et al. 1992; Lanier et al. 1992; Moretta et al. 1994). For example, mature NK cells can be obtained from in vitro $CD7^{+}CD34^{+}$ cells isolated from human thymus, when cultured in the presence of appropriate feeder cells or interleukin (IL)-15 (Mingari et al. 1997a). The molecular mechanism allowing NK cells to discriminate between normal and tumor cells, predicted by the "missing self hypothesis" (Ljunggren and Karre 1990), has been clarified only in recent years. NK

[1] Istituto Giannina Gaslini, Genova, Italy
[2] Dipartimento di Medicina Sperimentale, Università degli Studi di Genova, Italy
[3] Istituto Nazionale per la Ricerca sul Cancro, Genova, Italy
[4] Dipartimento di Oncologia, Biologia e Genetica, Università degli Studi di Genova, Italy

cells recognize MHC class I molecules through surface receptors delivering signals that inhibit, rather than activate NK cells. As a consequence, NK cells lyse target cells that have lost (or express insufficient amounts of) MHC class I molecules as frequently occurs in tumors and in virus-infected cells.

2 HLA Class I-Specific Inhibitory Receptors

Studies both in mouse and in man have revealed that NK cells are not equipped with a single type of receptor allowing detection of any MHC class I molecule. For, example, human NK cells express different inhibitory receptors, all characterized by immunoreceptor tyrosine-based inhibition motif (ITIM) in their cytoplasmic tail. While some of these receptors detect determinants shared by different alleles of a given HLA class I locus, others appear to display a broader specificity for various HLA class I molecules (Moretta et al. 1996; Lanier 1998). The first includes the so-called killer inhibitory receptors (KIRs) that are members of the Ig superfamily (IgSF) and are characterized by two or three Ig-like domains (Moretta et al. 1997). The second includes ILT-2/LIR-1 and p49 (Cosman et al. 1997; Samaridis and Colonna 1997; Cantoni et al. 1998a). Another important receptor is formed by the association between CD94 and NKG2A, both type II molecules containing a C-type lectin domain in their extracellular portion (Carretero et al. 1997). CD94/NKG2A molecules recognize HLA-E (Braud et al. 1998; Lee et al. 1998). This non-classical HLA class I molecule is surface expressed only in association with various HLA class I alleles belonging to different loci because it depends on the binding of peptides derived from the signal sequence of these HLA alleles (Braud et al. 1997; Borrego et al. 1998). As such, although specific for HLA-E, CD94/NKG2A operationally functions by sensing the overall expression of HLA class I molecules on potential target cells. Remarkably, each inhibitory receptor is expressed only by a fraction of NK cells. Moreover, although all NK cells express at least one receptor specific for self-MHC, the coexpression of two or more self-reactive receptors is uncommon in normal conditions (Moretta et al. 1996). This allows the whole NK cell pool of an individual to sense the loss of even single class I alleles on target cells. Thus, NK cells revealed a sophisticated capability of detecting potentially harmful cells. Table 1 illustrates the main HLA class I-specific inhibitory receptors expressed by human NK cells (see also Fig. 1).

Table 1. The HLA class I-specific inhibitory receptors expressed by human NK cells

NK receptor	Molecular family	Chromosomal localization	HLA-specificity
p58.1 (KIR2DL1)[a]	IgSF	19q13.4	HLA-Cw2, 4, 5, 6
p58.2 (KIR2DL2)[a]	IgSF	19q13.4	HLA-Cw1, 3, 7, 8
p70/NKB1 (KIR3DL1)[a]	IgSF	19q13.4	HLA-Bw4 alleles
p140 (KIR3DL2)[a]	IgSF	19q13.4	HLA-A3, A11...
CD94/NKG2A	C-type lectin	12p12–p13	HLA-E
ILT2/LIR1	IgSF	19q13.4	HLA-G and various HLA alleles
p49 (KIR2DL4)	IgSF	19q13.4	HLA-G and various HLA alleles

[a] In parentheses, another commonly used KIR denomination is indicated.

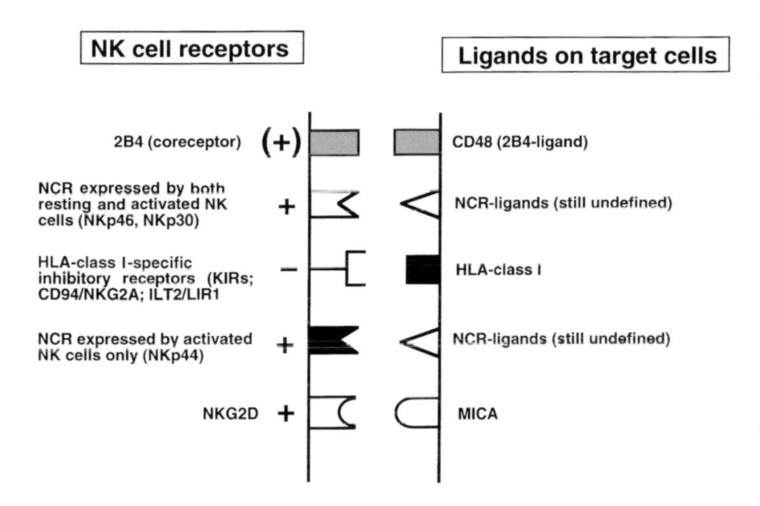

Fig. 1. Schematic representation of the surface receptors involved in NK cell activation or inactivation. The activating NK receptors interact with their ligands on target cells. In the absence of simultaneous inhibitory signals, e.g., in the case of HLA class I-negative target cells, this leads to NK cell triggering and target cell lysis. In the case of normal cells expressing HLA class I molecules, the interaction between inhibitory NK receptors and HLA class I molecules results in inhibitory signals which overcome NK cell triggering. The inhibitory receptors are heterogeneous in their molecular structure and HLA specificity (see Table 1). The activating receptors include constitutively expressed (NKp46 and NKp30) and inducible (NKp44) NCRs. The activating function of the 2B4 coreceptor is dependent on the simultaneous engagement of NCRs

3 Activating Receptors and Co-Receptors Involved in Natural Cytotoxicity

The fact that NK cells need to be inactivated in order to prevent lysis of normal MHC class I^+ cells together with the finding that they kill efficiently MHC class I-negative cells, clearly implies the existence of an "on" signal whenever NK cells interact with potential target cells.

Surface molecules expressed by NK cells that can mediate cell activation have been known for a long time. These include CD2 (Bolhuis et al. 1986), CD16 (Lanier et al. 1988) and CD69 (Moretta et al. 1991). However, none of these molecules appears to be directly involved in triggering of natural cytotoxicity. Studies suggest that, in humans, the activating forms of the HLA C-specific receptor (termed p50) (Moretta et al. 1995; Biassoni et al. 1996) and the CD94/NKG2C heterodimer (Cantoni et al. 1998b) play a role in the NK cytotoxicity against HLA class I^+ target cells. Both p50 and NKG2C receptors were found to be associated with a novel 12-kDa signal-transducing molecule that was termed KARAP/DAP-12 (Olcese et al. 1997; Lanier et al. 1998) and contains an immunoreceptor tyrosine-based activating motif (ITAM).

However, a major characteristic of NK cells is the efficient lysis of target cells that are deficient or completely lacking in HLA class I molecules (Moretta et al. 1992, 1996, 1997). Moreover, NK cell-mediated lysis of HLA class I^+ target cells may be induced by masking the HLA class I molecules on target cells with monoclonal antibodies (mAbs) (Ciccone et al. 1994; Moretta et al. 1997). These observations imply the existence of NK-cell receptors that are responsible for the induction of NK-cell triggering and which do not recognize HLA class I molecules. Indeed, different receptor/ligand interactions are probably responsible for NK-cell activation upon interaction with target cells. For example, individual NK-cell clones are heterogeneous in their ability to lyse different HLA class I-negative target cells. These data are in agreement with the concept that activation of NK cells in an HLA-independent context is the result of multiple interactions between NK receptors responsible for NK-cell triggering in the process of non-MHC-restricted natural cytotoxicity. Indeed, three novel NK-specific triggering surface molecules have been identified that appear to play a critical role in the induction of NK-mediated cytotoxicity (Sivori et al. 1997; Pessino et al. 1998; Vitale et al. 1998; Pende et al. 1999; Moretta et al. 2000). These molecules represent the first members of a novel emerging group of receptors that we termed "natural cytotoxicity receptors" (NCRs). NKp46, the first identified NCR, is expressed on both resting and activated NK cells and plays a key role in the lysis of various tumor cells. Its surface density (most normal donors have an $NKp46^{bright}$ phenotype) precisely correlates with the natural cytotoxicity of a given NK cell population or clone (Sivori et al. 1999). Remarkably, NKp46 is the only NCR involved in the recognition and lysis of murine targets by human NK cells. In agreement with this finding, an NKp46 homolog has been identified in mouse (Biassoni et al. 1999). NKp30 displays a pattern of surface expression overlapping with that of NKp46; however, the lysis of some tumors was found to be NKp46-independent and was selectively triggered via NKp30 (Pende et al. 1999). NKp44 is absent in resting NK cells but becomes expressed following NK cell activation in the presence of IL-2. The de novo expression of NKp44 upon activation is likely to be related, at least in part, to the higher efficiency of IL-2-cultured NK cells in mediating tumor cell lysis.

The molecular cloning of NKp46, NKp30, and NKp44 revealed novel members of the IgSF with no homology to each other and a low degree of homology with known human molecules. The cytoplasmic portion of NCR does not contain sequence motifs typically involved in the activation of the signal cascade(s). Thus, NCRs are coupled to the intracytoplasmic signal-transduction machinery with the ITAM-containing CD3ζ and/or FcεRγ (NKp46 and NKp30) or KARAP/DAP12 (NKp44) adapter polypeptides. Although the NK-mediated lysis of the majority of tumor-cell lines analyzed is dependent upon the NCR identified so far, experimental data suggest the existence of other receptors playing a role in the cytolysis activity against certain tumors (e.g., some T cell leukemias). A suitable candidate for this function appeared to be the 2B4 molecule which had been shown to trigger the NK cytotoxicity in both mouse (Garni-Wagner et al. 1993; Mathew et al. 1993) and human (Valiante et al. 1993; Braun et al. 1998; Latchman et al. 1998; Kubin et al. 1999; Nakajima et al. 1999; Sivori et al. 2000). Recent data indicated that 2B4 functions as a co-receptor rather than as a true receptor. Thus, the differential ability of NK cells to respond to 2B4 was found to strictly correlate with their NCR phenotype. While the $NKp46^{bright}$ NK cell clones could be triggered by anti-2B4 mAb, the $NKp46^{dull}$ could not, in spite of a similar surface density of 2B4. Moreover, mAb induced modulation of NKp46 prevented NK cells from subsequently being triggered by anti-2B4 mAb (Sivori et al. 2000).

A possible role in NK-mediated cytotoxicity has recently been suggested also for NKG2D, a receptor for the stress-inducible MICA (broadly expressed on tumors of epithelial origin). NKG2D, different from NCR, is expressed also by cytolytic T cells and has been reported to overrule the inhibitory signals delivered by the KIR/HLA-class I interactions (Bauer et al. 1999).

Regarding the ligands recognized on target cells by the various triggering receptors or coreceptors involved in NK-mediated cytotoxicity, they are known in the case of (1) the activating counterparts of the inhibitory NK receptors (Moretta et al. 1996; Biassoni et al. 1997; Moretta et al. 1997) including p50.1, p50.2 and NKG2C (i.e. HLA-class I molecules)

(Moretta et al. 1996, 1997; Cantoni et al. 1998b), (2) NKG2D (i.e., the stress-inducible MICA) (Bauer et al. 1999) and (3) 2B4 (i.e. CD48) (Braun et al. 1998; Latchman et al. 1998; Kubin et al. 1999; Nakajima et al. 1999). The molecular nature of the NCR-specific ligands has still to be defined. However, experimental evidence suggests that they are expressed both by normal and tumor cells.

4 Expression of HLA Class I-Specific Inhibitory Receptors in Cytolytic T Lymphocytes

Inhibitory receptors, originally identified in NK cells were also detected in other leukocyte types, thus indicating that they may play a more general role in the control of leukocyte function. For example, the HLA class I-specific inhibitory receptors were found to be expressed by some activated cytolytic T lymphocytes (CTLs) in which they can inhibit TCR-mediated T cell activation and function (Mingari et al. 1995, 1998a).

All of the known inhibitory NKRs can be expressed by T cells (Mingari et al. 1997a), primarily by the $CD8^+$ subset. In addition, analysis of markers informative of the memory/naïve T-cell phenotype revealed that NKR^+ T cells lack CD28, while they express high levels of CD18, CD29, CD57, and CD45RO (Mingari et al. 1996). These data, together with the finding that T cells expressing NKR are absent in thymus and cord blood (Mingari et al. 1997a,b), are consistent with a memory phenotype and indicate that they represent cells activated in vivo. In normal donors, NKR^+ T cells express only one or two TCRVβ (different in different individuals); moreover, each Vβ family analyzed exhibited identical rearrangements (Mingari et al. 1996). Therefore, NKR^+ T cells in normal donors appear to represent oligoclonal or monoclonal T cell expansions. It is conceivable that they may be the result of a chronic, antigen-driven stimulation.

A possible explanation for the advantage of NKR expression may be offered by those CTLs that have acquired NK-like activity and would thus be deleterious to normal cells, if they did not express NKR (see below). The expression of NKR can also be viewed as a regulatory mechanism useful for the fine-tuning of TCR-mediated responses. In this context, NKR expression might play a physiological role in preventing CTLs from cross-reacting with self-antigens, while their defective expression could be involved in autoimmune diseases caused by autoreactive CTLs.

While NKR expression may prevent killing of normal cells by CTLs with NK-like activity, the other side of the coin is the inability of these cells to properly control viral infections or tumor growth, at least in the case of target cells expressing MHC-class I molecules. In favor of this possibility are data in tumor-bearing (IKEDA et al. 1997) or in immunodeficiency virus (HIV)-infected patients (DE MARIA et al. 1997). A tumor-specific (HLA-A24-restricted) CTL clone isolated from a melanoma patient expressed p58.2 molecules (IKEDA et al. 1997). This clone did not lyse the original autologous tumor (expressing the complete set of HLA class I alleles), while it lysed an HLA-loss variant of the tumor that expressed HLA-A24 but lost all of the other class I alleles, including HLA-Cw7, which is a ligand for p58.2 KIR (IKEDA et al. 1997). Regarding viral diseases, high proportions of NKR$^+$ T cells are frequently detectable in HIV-infected patients (DE MARIA et al. 1997). More importantly, KIRs were detected in HIV-specific CTLs and inhibited lysis of autologous lymphoblastoid cells expressing HIV antigens (DE MARIA et al. 1997). MAb-mediated masking of KIRs restored target cell lysis, thus indicating that KIR$^+$ CTLs are not anergic but are simply inhibited in their function as a consequence of the KIR/class I interaction. It is of note that KIR crosslinking also inhibits cytokine (e.g., IFN-γ) production by CTLs, thus suggesting an interference with different defense mechanisms mediated by CTLs (KAGI and HENGARTNER 1996).

These data underscore the importance of identifying the mechanism(s) leading to the expression of KIRs in T lymphocytes (MINGARI et al. 1998a). We found that T lymphocytes responding in vitro to superantigens or allogeneic cells in the presence of IL-15 expressed CD94/NKG2A. This expression was mostly confined to CD8$^+$ cells and, in alloreactive CTLs, led to an impairment of allospecific cytolytic activity that could be reversed by mAb-mediated masking of the receptor (MINGARI et al. 1998b).

We analyzed whether other cytokines displayed a similar effect. We speculated that the induction of NKRs could represent one of the mechanisms utilized by inhibitory cytokines to downregulate immune responses. Indeed, transforming growth factor-β (TGF-β) was found to induce CD94/NKG2A expression in superantigen-stimulated T cells in vitro (BERTONE et al. 1999).

How do data on the effect of cytokines fit with the detection of KIRs in CTLs directed to tumor or viral antigens? It should be noted that tumors may produce cytokines including TGF-β. In addition, evidence has been obtained that melanoma cells may also synthetize IL-15 (MEAZZA et al. 1997). Therefore, the tumor itself could induce the expression of NKR in tumor-specific CTLs. This would represent yet another mechanism of tumor escape from CTL-mediated control. Similarly, viral infection may result in the release of cytokines, (from infected cells or as a result of the inflammatory response) which could influence KIR expression.

5 Inhibitory NK Receptors Expressed in Myeloid Cells

Inhibitory receptors of still-undefined specificity, originally identified in NK cells, are expressed in monocytes, dendritic cells (DC), or granulocytes in which they may mediate inhibition of cell function or proliferation (Cella et al. 1997; Ulyanova et al. 1999; Vitale et al. 1999; Ferlazzo et al. 2000). For example, p75/AIRM1, a recently identified member of the sialoadhesin family displaying homology with the major myeloid cell marker CD33 (Falco et al. 1999; Nicoll et al. 1999) functions as a receptor capable of blocking proliferation of normal or leukemic myeloid cells (Vitale et al. 1999). Remarkably, on the basis of these findings, CD33 has also been carefully reinvestigated. These studies allowed assigning to this molecule a similar inhibitory function. For example, CD33 could inhibit the in vitro proliferation of CML and AML and could also block maturation of monocytes or $CD34^{+}$ cells towards DC (Vitale et al. 1999; Ferlazzo et al. 2000; Vitale et al. 2001). Another inhibitory receptor belonging to the ILT/LIR1 family (ILT3) could negatively regulate the activation of antigen-presenting cells (APCs) (Cella et al. 1997). Preliminary analyses of the mechanism(s) by which p75, CD33, and p40 (another inhibitory receptor shared by NK and myeloid cells) (Poggi et al. 1998) block cell proliferation would indicate the induction of apoptosis. This suggests that they may represent surface receptors initiating novel apoptotic pathways. Since, different from other surface receptors mediating apoptosis, both p75 and CD33 display a restricted pattern of tissue distribution, they could represent useful therapeutic tools to specifically target macrophages or DC and inhibit their maturation or function (e.g., in the case of diseases characterized by an overwhelming immune response or macrophage activation). In addition, they may be used to stop proliferation of myeloid leukemic cells. Thus, novel inhibitory receptors originally identified in NK cells are now becoming of more general interest. Indeed they appear to play a relevant regulatory in the development and function of DC, known to play a key role in the immune response, as well as in the process of normal or leukemic myelopoiesis.

Acknowledgements. This work was supported by grants awarded by Associazione Italiana per la Ricerca sul Cancro (AIRC), Istituto Superiore di Sanità (ISS), Ministero della Sanità, and Ministero dell'Università e della Ricerca Scientifica e Tecnologica (MURST), Consiglio Nazionale delle Ricerche, Progetto Finalizzato Biotecnologie, Biotechnology Program L.95/95). The financial support of Telethon-Italy (grant no. E.0892) is gratefully acknowledged.

References

Bauer S, Groh V, Wu J, Steinle A, Phillips, JH, Lanier LL, Spies T (1999) Activation of NK cells and T cells by NKG2D, a receptor for stress-inducible MICA. Science 285:727–729

Bertone S, Schiavetti F, Bellomo R, Vitale C, Ponte M, Moretta L, Mingari MC (1999) Transforming growth factor-β-induced expression of CD94/NKG2A inhibitory receptors in human T lymphocytes. Eur J Immunol 29:23–29

Biassoni R, Pessino A, Bottino C, Pende D, Moretta L, Moretta A (1999) The murine homologue of the human NKp46, a triggering receptor involved in the induction of natural cytotoxicity. Eur J Immunol 29:1014–1020

Biassoni R, Pessino A, Malaspina A, Cantoni C, Sivori S, Moretta L, Moretta A (1997) Role of amino acid position 70 in the binding affinity of p50.1 and p58.1 receptors for HLA-Cw4 molecules. Eur J Immunol 27:3095–3099

Biassoni R, Cantoni C, Falco M, Verdiani S, Bottino C, Vitale M, Conte R, Poggi A, Moretta A, Moretta L (1996) The human Leukocyte Antigen (HLA)-C-specific "activatory" or "inhibitory" Natural Killer cell receptors display highly homologous extracellular domains but differ in their transmembrane and intracytoplasmic portions. J Exp Med 183:645–650

Bolhuis RLH, Roozemond RC, Van de Griend RJ (1986) Induction and blocking of cytolysis in $CD2^+$, $CD3^-$ NK and $CD2^+$, $CD3^+$ cytotoxic T lymphocytes via CD2 50 KD sheep erythrocyte receptor. J Immunol 136:3939–3944

Borrego F, Ulbrecht M, Weiss EH, Coligan JE, Brooks AG (1998) Recognition of Human Histocompatibility Leukocyte Antigen (HLA)-E complexed with HLA class I signal sequence-derived peptides by CD94/NKG2 confers protection from Natural Killer cell-mediated lysis. J Exp Med 187:813–818

Braud VM, Jones EY, McMichael AJ (1997) The human major histocompatibility complex class Ib molecule HLA-E binds signals-derived peptides with primary anchor residues at position 2 and 9. Eur J Immunol 27:1164–1169

Braud VM, Allan DSJ, O'Callaghan CA, Soderstrom K, D'Andrea A, Ogg GS, Lazetic S, Young NT, Bell JI, Phillips JH, Lanier LL, McMichael AJ (1998) HLA-E binds to natural killer cell receptors CD94/NKG2A, B and C. Nature 391:795–799

Brown MH, Boles K, van der Merwe PA, Kumar V, Mathew PA, Barclay AN (1998) 2B4, the natural killer and T cell immunoglobulin superfamily surface protein, is a ligand for CD48. J Exp Med 188:2083–2090

Cantoni C, Verdiani S, Falco M, Pessino A, Cilli M, Conte R, Pende D, Ponte M, Mikaelsson MS, Moretta L, Biassoni R (1998a) p49, a novel putative HLA-class I-specific inhibitory NK receptor belonging to the Immunoglobulin Superfamily. Eur J Immunol 28:1980–1990

Cantoni C, Biassoni R, Pende D, Sivori S, Accame L, Pareti L, Semenzato G, Moretta L, Moretta A, Bottino C (1998b) The activating form of CD94 receptor complex. The CD94-associated Kp39 protein represents the product of the NKG2-C gene. Eur J Immunol 28:327–338

Carretero M, Cantoni C, Bellon T, Bottino C, Biassoni R, Rodriguez A, Perez-Villar JJ, Moretta L, Moretta A, Lopez-Botet M (1997) The CD94 and NKG2-A C-type lectins covalently assemble to form a NK cell inhibitory receptor for HLA class I molecules. Eur J Immunol 27:563–567

Cella M, Dohring C, Samaridis J, Dessing M, Brockhaus M, Lanzavecchia A, Colonna MJ (1997) A novel inhibitory receptor (ILT3) expressed on monocytes, macrophages, and dendritic cells involved in antigen processing. J Exp Med 185:1743–1751

Ciccone E, Pende D, Vitale M, Nanni N, Di Donato C, Bottino C, Morelli L, Viale L, Amoroso A, Moretta A, Moretta L (1994) Self Class I molecules protect normal cells from lysis mediated by autologous Natural Killer cells. Eur J Immunol 24:1003–1006

Cosman D, Fanger N, Borges L, Kubin M, Chin W, Peters L, Hsu M (1997) A novel immunoglobulin superfamily receptor for cellular and viral MHC class I molecules. Immunity 7:273–282

De Maria A, Ferraris A, Guastella M, Pilia S, Cantoni C, Polero L, Mingari MC, Bassetti D, Fauci AS, Moretta L (1997) Expression of HLA class I-specific inhibitory natural killer cell receptors in HIV-specific cytolytic T lymphocytes: impairment of specific cytolytic functions. Proc Natl Acad Sci USA 94:10285–10288

Falco M, Biassoni R, Bottino C, Vitale M, Sivori S, Augugliaro R, Moretta L, Moretta A (1999) Identification and molecular cloning of p75/AIRM1, a novel member of the sialoadhesin family which functions as an inhibitory receptor in human natural killer cells. J Exp Med 190:793–801

Ferlazzo G, Spaggiari GM, Semino C, Melioli G, Moretta L (2000) Engagement of CD33 surface molecules prevents the generation of dendritic cells from both monocytes and CD34$^+$ myeloid precursors. Eur J Immunol 30:827–833

Garni-Wagner BA, Purohit A, Mathew PA, Bennett M, Kumar V (1993) A novel function-associated molecule related to non-MHC-restricted cytotoxicity mediated by activated natural killer cells and T cells. J Immunol 151:60–70

Ikeda H, Lethé B, Lehmann F, Van Baren N, Baurain JF, De Smet C, Chambost H, Vitale M, Moretta A, Boon T, Coulie PG (1997) Characterization of an antigen that is recognized on a melanoma showing partial HLA loss by CTL expressing an NK inhibitory receptor. Immunity 6:199–208

Kagi D, Hengartner H (1996) Different roles for cytotoxic T cells in the control of infections with cytopathic versus noncytopathic viruses. Curr Opin Immunol 8:472–477

Kubin MZ, Parsley DL, Din W, Waugh JY, Davis-Smith T, Smith CA, Macduff BM, Armitage RJ, Chin W, Cassiano L, Borges L, Petersen L, Trinchieri G, Goodwing RG (1999) Molecular cloning and biological characterization of NK cell activation-inducing ligand, a counterstructure for CD48. Eur J Immunol 29:3466–3477

Lanier LL (1998) NK cell receptors. Annu Rev Immunol 16:359–393

Lanier LL, Ruitenberg JJ, Phillips JH (1988) Functional and biochemical analysis of CD16 antigen on natural killer cells and granulocytes. J Immunol 141:3478–3485

Lanier LL, Spits H, Phillips JH (1992) The developmental relationship between NK cells and T cells. Immunol Today 13:392–395

Lanier LL, Corliss BC, Wu J, Leong C, Phillips JH (1998) Immunoreceptor DAP12 bearing a tyrosine-based activation motifs is involved in activating NK cells. Nature 391:703–707

Latchman Y, McKay PF, Reiser H (1998) Identification of the 2B4 molecule as a counter-receptor for CD48. J Immunol 161:5809–5812

Lee N, Llano M, Carretero M, Ishitani A, Navarro F, Lòpez-Botet M, Gherarty D (1998) HLA-E is a major ligand for the natural killer inhibitory receptor CD94/NKG2A. Proc Natl Acad Sci USA 95:5199–5204

Ljunggren HG, Karre K (1990) In search of the "missing self". MHC molecules and NK cell recognition. Immunol Today 11:237–241

Mathew PA, Garni-Wagner BA, Land K, Takashima A, Stoneman E, Bennett M, Kumar V (1993) Cloning and characterization of the 2B4 gene encoding a molecule associated with non-MHC-restricted killing mediated by activated natural killer cells and T cells. J Immunol 151:5328–5337

Meazza R, Gaggero A, Neglia F, Basso S, Sforzini S, Pereno R, Azzarone B, Ferrini S (1997) Expression of two interleukin-15 mRNA isoforms in human tumors does not correlate with secretion: role of different signal peptides. Eur J Immunol 27:1049–1054

Mingari MC, Moretta A, Moretta L (1998a) Regulation of KIR expression in human T lymphocytes. A safety mechanism which may impair protective T cell responses. Immunol Today 19:153–157

Mingari MC, Ponte M, Bertone S, Schiavetti F, Vitale C, Bellomo R, Moretta A, Moretta L (1998b) HLA class I-specific inhibitory receptors in human T lymphocytes: interleukin

15-induced expression of CD94/NKG2A in superantigen- or alloantigen-activated $CD8^+$ T cells. Proc Natl Acad Sci USA 95:1172–1177
Mingari MC, Poggi A, Biassoni R, Bellomo R, Ciccone E, Pella N, Morelli L, Verdiani S, Moretta A, Moretta L (1991) In vitro proliferation and cloning of CD3-$CD16^+$ cells from human thymocyte precursors. J Exp Med 174:21–26
Mingari MC, Vitale C, Cambiaggi A, Schiavetti F, Melioli G, Ferrini S, Poggi A (1995) Cytolytic T lymphocytes displaying Natural Killer (NK)-like activity. Expression of NK-related functional receptors for HLA class I molecules (p58 and CD94) and inhibitory effect on the TCR-mediated target cell lysis or lymphokine production. Int Immunol 7:697–703
Mingari MC, Schiavetti F, Ponte M, Vitale C, Maggi E, Romagnani S, Demarest J, Pantaleo G, Fauci AS, Moretta L (1996) Human $CD8^+$ T lymphocyte subsets that express HLA class I-specific inhibitory receptors represent oligoclonally or monoclonally expanded cell population. Proc Natl Acad Sci USA 93:12433–12438
Mingari MC, Ponte M, Cantoni C, Vitale C, Schiavetti F, Bertone S, Bellomo R, Tradori-Cappai A, Biassoni R (1997a) HLA-class I-specific inhibitory receptors in human cytolytic T lymphocytes: molecular characterization, distribution in lymphoid tissues and co-expression by individual T cells. Int Immunol 9:485–491
Mingari MC, Vitale C, Cantoni C, Bellomo R, Ponte M, Schiavetti F, Bertone S, Moretta A, Moretta L (1997b) Interleukin-15-induced maturation of human Natural Killer cells from early thymic precursors. Selective expression of CD94/NKG2A as the only HLA-class I specific inhibitory receptor. Eur J Immunol 27:1374–1380
Moretta L, Ciccone E, Mingari MC, Biassoni R, Moretta A (1994) Human NK cells: origin, clonality, specificity and receptors. Adv Immunol 55:341–380
Moretta A, Biassoni R, Bottino C, Mingari MC, Moretta L (2000) The natural cytotoxicity receptors that trigger the human NK-mediated cytolysis. Immunol Today 21:228–234
Moretta L, Ciccone E, Moretta A, Hoglund P, Ohlen C, Karre K (1992) Allorecognition by NK cells: nonself or no self? Immunol Today 13:300–306
Moretta A, Bottino C, Vitale M, Pende D, Biassoni R, Mingari MC, Moretta L (1996) Receptors for HLA-class I-molecules in human Natural Killer cells. Annu Rev Immunol 14:619–648
Moretta A, Sivori S, Vitale M, Pende D, Morelli L, Augugliaro R, Bottino C, Moretta L (1995) Existence of both inhibitory (p58) and activatory (p50) receptors for HLA-C molecules in human Natural Killer cells. J Exp Med 182:875–884
Moretta A, Biassoni R, Bottino C, Pende D, Vitale M, Poggi A, Mingari MC, Moretta L (1997) Major histocompatibility complex class I-specific receptors on human Natural Killer and T lymphocytes. Immunol Rev 155:105–117
Moretta A, Poggi A, Pende D, Tripodi G, Orengo AM, Pella N, Augugliaro R, Bottino C, Ciccone E, Moretta L (1991) CD69-mediated pathway of lymphocyte activation. Anti-CD69 mAbs trigger the cytolytic activity of different lymphoid effector cells with the exception of cytolytic T lymphocytes expressing TCR α/β. J Exp Med 174:1393–1398
Nakajima H, Cella M, Langen H, Friedlein A, Colonna M (1999) Activating interactions in human NK cell recognition: the role of 2B4-CD48. Eur J Immunol 29:1676–1683
Nicoll G, Ni J, Liu D, Klenerman P, Munday J, Dubock S, Mattei M-G, Crocker PR (1999) Identificaiton and characterization of a novel siglec, siglec-7, expressed by human natural killer cells and monocytes. J Biol Chem 274:34089–34095
Olcese L, Cambiaggi A, Semenzato G, Bottino C, Moretta A, Vivier E (1997) Human killer cell activatory are included in a multimeric complex expressed by Natural Killer cells. J Immunol 158:5083–5086
Pende D, Parolini S, Pessino A, Sivori S, Augugliaro R, Morelli L, Marcenaro E, Accame L, Malaspina A, Biassoni R, Bottino C, Moretta L, Moretta A (1999) Identification and molecular characterization of NKp30, a novel triggering receptor involved in natural cytotoxicity mediated by human natural killer cells. J Exp Med 190:1505–1516

Pessino A, Sivori S, Bottino C, Malaspina A, Morelli L, Moretta L, Biassoni R, Moretta A (1998) Molecular cloning of NKp46: a novel member of the immunoglobulin superfamily involved in triggering of Natural cytotoxicity. J Exp Med 188:953–960
Poggi A, Tomasello E, Ferrero E, Zocchi MR, Moretta L (1998) p40/LAIR-1 regulates the differentiation of peripheral blood precursors to dendritic cells induced by granulocyte–monocyte colony stimulating factor. Eur J Immunol 28:2086–2091
Rodewald HR, Moingeon P, Lucich JL, Dosiou C, Lopez P, Reinherz EL (1992) A population of early fetal thymocytes expressing FcγRII/III contains precursors of T lymphocytes and natural killer cells. Cell 69:139–150
Samaridis J, Colonna M (1997) Cloning of novel immunoglobulin superfamily receptors expressed on human myeloid and lymphoid cells: structural evidence for new stimulatory and inhibitory pathways. Eur J Immunol 2:660–665
Sivori S, Vitale M, Morelli L, Sanseverino L, Augugliaro R, Bottino C, Moretta L, Moretta A (1997) p46, a novel natural killer cell-specific surface molecule which mediates cell activation. J Exp Med 186:1129–1136
Sivori S, Pende D, Bottino C, Marcenaro E, Pessino A, Biassoni R, Moretta L, Moretta A (1999) NKp46 is the major triggering receptor involved in the natural cytotoxicity of fresh of cultured human NK cells. Correlation between surface density of NKp46 and natural cytotoxicity against autologous, allogeneic or xenogeneic target cells. Eur J Immunol 29:1656–1666
Sivori S, Parolini S, Falco M, Marcenaro E, Biassoni R, Bottino C, Moretta L, Moretta A (2000) 2B4 functions a co-receptor in human Natural Killer cell activation. Eur J Immunol 30:787–793
Trinchieri G (1989) Biology of natural killer cells. Adv Immunol 47:187–376
Ulyanova T, Blasioli J, Woodford-Thomas TA, Thomas ML (1999) The sialoadhesin CD33 is myeloid-specific inhibitory receptor. Eur J Immunol 29:2440–3449
Valiante NM, Trinchieri G (1993) Identification of a novel signal transduction surface molecule on human cytotoxic lymphocytes. J Exp Med 178:1397–1406
Vitale C, Romagnani C, Falco M, Vitale M, Moretta A, Bacigalupo A, Moretta L, Mingari MC (1999) Engagement of p75/AIRM1 or CD33 inhibits the proliferation of normal or leukemic myeloid cells. Proc Natl Acad Sci USA 96:15091–15096
Vitale M, Bottino C, Sivori S, Sanseverino L, Castriconi R, Marcenaro E, Augugliaro R, Moretta L, Moretta A (1998) NKp44, a novel triggering surface molecule specifically expressed by activated Natural Killer cells is involved in non-MHC restricted tumor cell lysis. J Exp Med 187:2065–2072
Vitale C, Romagnani C, Puccetti A, Olive D, Costello R, Chiossone L, Pitto A, Bacigalupo A, Moretta L, Mingari MC (2001) Surface expression and function of p75/AIRM1 or CD33 in acute myeloid leukemias. Engagement of CD33 induces apoptosis of leukemic cells. Proc Natl Acad Sci (USA) 98:5764–5769

Checkpoints in the Regulation of T Helper 1 Responses

V.L. Heath, H. Kurata, H.J. Lee, N. Arai, and A. O'Garra

1 Introduction

The immune system has developed to be highly specialized and effective in eradicating a wide variety of pathogens with a minimum of immunopathology. The adaptive arm of the immune response, consisting of antigen-specific T and B cells, interacts with cells of the innate immune system to mediate an effective response to infectious pathogens. Heterogeneity of T cell responses to pathogens can determine the resistance or susceptibility to such infectious agents. In this regard, at least two subsets of $CD4^+$ T helper subsets have been identified; these subsets are characterized by the cytokines they produce and play distinct roles in fighting infection as well as contributing to immunopathology (Mosmann et al. 1986; Romagnani 1991; Sher and Coffman 1992; Abbas 1996). T helper (Th)1 cells produce interferon (IFN)-γ and lymphotoxin and play a critical role in cell-mediated immunity (Mosmann et al. 1986; Sher and Coffman 1992). These cells have also been implicated as being involved in organ-specific autoimmune diseases, such as multiple sclerosis and insulin-dependent diabetes, as well

Department of Immunobiology, DNAX Research Institute, 901 California Avenue, Palo Alto, CA 94304, USA as chronic inflammatory diseases including inflammatory bowel disease (Powrie and Coffman 1993; Liblau et al. 1995; O'Garra 1998). Th2 cells produce interleukin (IL)-4, IL-5, IL-10, and IL-13 and are important in helminth immunity, as well as playing a role in allergy and atopy (Romagnani 1994; O'Garra 1998). Gaining an understanding of the molecular mechanisms of Th1 and Th2 development is critical both for the development of more powerful anti-microbial agents as well as for the development of therapies for immunopathologies.

We and others have investigated the molecular mechanisms of Th1 development, both in terms of soluble factors which drive Th1 development and the signaling pathways and transcription factors that are involved. Two main systems are used for these types of studies. The first is the polyclonal stimulation of T cells with, for example, anti-CD3 and IL-2. The second system makes use of T cells derived from mice expressing transgenic T cell receptors of known specificities which can be activated in an antigen-specific manner. These systems have proved invaluable for determining which factors are required for Th1 and Th2 development, and importantly it has been shown that both Th1 and Th2 cells can be derived from the same precursor cell (Rocken 1992; Kamogawa et al. 1993). IL-12 is dominant in driving the development of Th1 cells which produce IFN-γ (Hsieh et al. 1993; Manetti et al. 1993; Seder et al. 1993; Trinchieri 1995). IL-12 consists of two subunits, p35 and p40, and upon binding to its cognate receptor, consisting of the IL-12 receptor(R)β1 and the IL-12Rβ2 chains, activates signal transducer and activator of transcription (STAT) proteins 1, 3, and 4 (Bacon et al. 1995; Jacobson et al. 1995; Szabo et al. 1995, 1997; Presky et al. 1996; Rogge et al. 1997). In contrast to Th1 cells, Th2 cells were shown to lack IL-12-mediated activation of STAT3 and STAT4 (Szabo et al. 1995, 1997; Hilkens et al. 1996). The molecular basis for this was found to be that Th2 cells downregulated the IL-12Rβ2 chain and so became unresponsive to IL-12 (Rogge et al. 1997; Szabo et al. 1997). Targeted disruption of IL-12 p40 (Magram et al. 1996), IL-12Rβ1 (Wu et al. 1997a), and STAT4 (Thierfelder et al. 1996) each resulted in mice with a reduced Th1 response and impaired cell-mediated immunity.

2 The Role of IL-1α and IL-18 in Th Cell Responses

In addition to IL-12, we have found other cytokines that can potentiate Th1 development. We consistently observed that $CD4^+$ Th cells from BALB/c mice, developed in cultures using low numbers of purified dendritic cells and IL-12, produced significantly less IFN-γ than similar cells developed in cultures using unfractionated splenic antigen-presenting cells (APC) and IL-12 (Shibuya et al. 1998) (Fig. 1A,B). We reasoned that this missing link might be due to other cytokines produced by the splenocytes which augment the action of IL-12. In a series of experiments we found that IL-1α alone, and to a greater extent IL-1α in combination with tumor necrosis factor (TNF)-α, could act as cofactors with IL-12 to induce the development of Th1 cells that produce increased amounts of IFN-γ

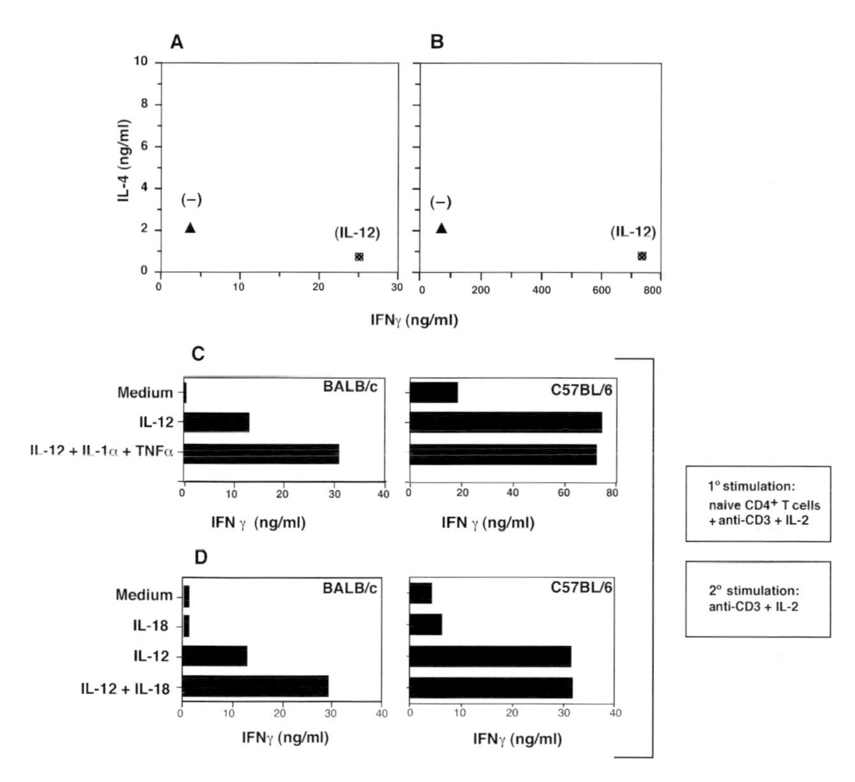

Fig. 1. A,B IL-12-driven Th1 development upon stimulation with spleen APC results in significantly higher levels of IFN-γ production by Th1 cells than stimulation with dendritic cells. Naïve $CD4^+$ T cells from DO11.10 T cell receptor (TCR) transgenic mice (BALB/c background) were cultured with irradiated fluorescence-activated cell sorter (FACS) sorted dendritic cells (**A**) or irradiated whole spleen (**B**) and antigen in medium alone (*triangles*) or with IL-12 (*squares*). Seven days after the initiation of the culture cells were harvested and restimulated at equal cell density with splenic APC and antigen. Supernatants were harvested at 48h and IFN-γ and IL-4 were measured by enzyme-linked immunosorbent assay (ELISA). **C,D** IL-1α plus TNF-α and IL-18 potentiate IL-12-induced Th1 development from BALB/c but not C57BL/6 T cells stimulated with anti-CD3 and IL-2. Naïve $CD4^+$ T cells were sorted from BALB/c and C57BL/6 mice, cultured with crosslinked anti-CD3, IL-2 and medium or the cytokine combinations as noted on the figure. After 7 days the cells were harvested and restimulated at equal cell number with anti-CD3 and IL-2. Supernatants were harvested at 48h and IFN-γ measured by ELISA

(Fig. 1C). Interestingly these cofactors for IL-12-driven Th1 development did not augment the IFN-γ production by T cells from C57BL/6 mice, which produced maximal IFN-γ when cultured with dendritic cells and IL-12. The molecular basis for this genetic difference between these strains is not clear, although we found that TNF-α and IL-1α were required for maximal expression of the IL-2Rα (CD25) by Th1 cells in BALB/c but not C57BL/6 mice. Previously IL-1α had not been shown to play a role in the development of Th1 cells. This cytokine was originally defined as a factor that promoted the proliferation of murine thymocytes (Gery and Lepe-Zuniga 1983). More recently, IL-1α was shown to increase IL-4 responsiveness of Th2 clones and thus enhance their proliferation, whereas Th1 clones were shown to be unresponsive to IL-1α (Greenbaum et al. 1988; Lichtman et al. 1988; McArthur and Raulet 1993). Both TNF-α and IL-1α have been shown to modestly upregulate IL-12 binding to resting peripheral blood mononuclear cells (PBMC), and so it is possible that these cytokines might upregulate the IL-12R on T cells from BALB/c mice, thus increasing their IL-12 responsiveness (Wu et al. 1997b).

We had observed a second missing link in that developing and committed T cell receptor (TCR) transgenic Th1 cells – from either C57BL/6 or BALB/c mice, restimulated with anti-CD3 and IL-2, even in the presence of IL-12 – produced significantly less IFN-γ than when they were restimulated with splenic antigen presenting cell (APC) and antigen. In this situation we found that IL-1α was not a cofactor for IL-12-induced IFN-γ production, in keeping with previous observations as previously discussed (Greenbaum et al. 1988; Lichtman et al. 1988). Rather we found another cytokine, IL-18, to be important in potentiating IFN-γ production by Th1 cells. This factor was originally described as a serum factor induced by infection with *P. acnes* and challenge with lipopolysaccharide (LPS) which could stimulate IFN-γ production by T cells and NK cells, and named interferon gamma inducing factor (IGIF) (Okamura et al. 1995). Upon more detailed structural analysis, this cytokine was identified as being a member of the IL-1 family (designated IL-1γ) (Bazan et al. 1996). Interestingly, in common with IL-1α, IL-18 was found to potentiate IL-12-driven Th1 development of T cells from BALB/c, but not C57BL/6 mice (Robinson 1997) (Fig. 1D). However, IL-1α and IL-18 were found to have very different effects on fully differentiated Th1 and Th2 cells. IL-18 was found to act on differentiated Th1 cells where it synergized with IL-12 to induce maximal IFN-γ production, whereas IL-1α did not (Robinson 1997) (Fig. 2A). In contrast, IL-1α induced proliferation of Th2 cells, whereas IL-18 affected neither proliferation nor cytokine production by this Th subset (Fig. 2B). It was of interest to observe that IL-1α and IL-18 activated IL-1

A Three-week polarized Th1 cells

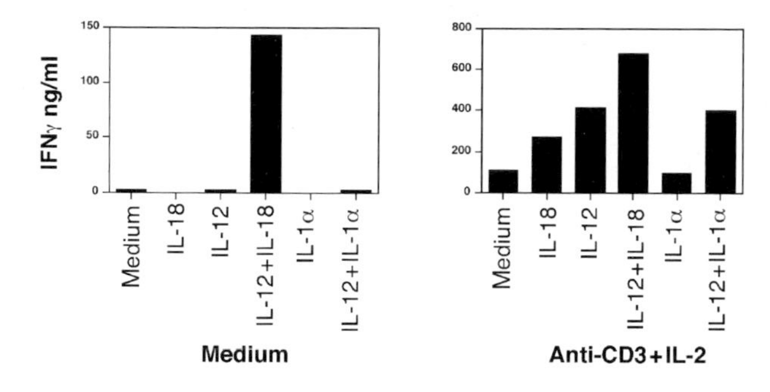

B Three-week polarized Th2 cells

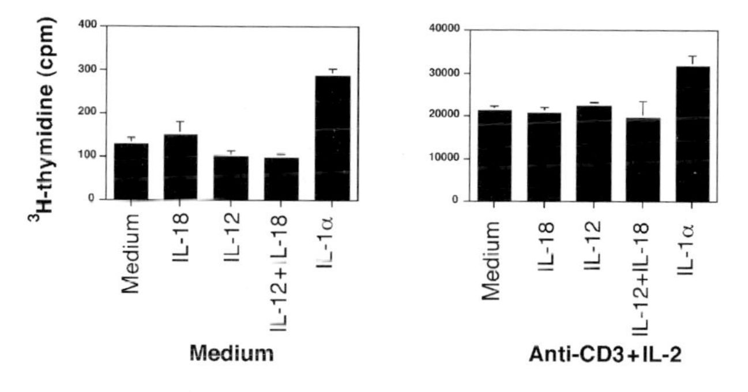

Fig. 2. IL-18 potentiates IL-12-induced proliferation and IFN-γ production from committed Th1 cells but does not affect Th2 cells which respond to IL-1α. **A**,**B** Committed Th1 and Th2 cells were derived from na $CD4^+$ T cells from TCR7 TCR transgenic mice (C57BL/6 background) stimulated with antigen presented by irradiated splenic APC for three rounds of 1 week in the presence of IL-12 and anti-IL-4 (Th1 cells) or IL-4 and anti-IL-12 (Th2 cells). Cells were rested in IL-2 prior to the assay. Th1 and Th2 cells were treated in either medium alone or with soluble anti-CD3 and IL-2 with cytokines added as noted in the figure. In the case of Th1 cells (**A**) supernatants were harvested at 48h and IFN-γ measured by ELISA. Proliferation of Th2 cells (**B**) was assessed using [^{3}H]thymidine incorporation

receptor-associated kinase (IRAK) and induced the nuclear translocation of NFκBp50/p65 exclusively in Th2 or Th1 cells, respectively. The differential activation of a common signaling pathway by different IL-1 family members in either Th1 or Th2 cells strongly suggested that these cytokines bound to distinct and differentially regulated receptors. The IL-1α receptor consists of the type I IL-1R and the IL-1R accessory protein (IL-1RAcP) (Sims et al. 1988; Greenfeder et al. 1995). IL-1α/β receptors were shown to be lost upon differentiation to the Th1 phenotype (Lichtman et al. 1988). Antibodies to the type I IL-1R, the type II IL-1R, and the IL-1RAcP did not block the activity of IL-18 (Hunter et al. 1997). More recently the

IL-18 receptor has been characterized. One chain designated IL-1R related protein (IL-1Rrp) was first identified as an orphan receptor of the IL-1R family (Parnet et al. 1996), and this was subsequently found to be a receptor for IL-18 (Torigoe et al. 1997; Xu et al. 1998). Consistent with the activity of IL-18 on differentiated Th1 but not Th2 cells, IL-1Rrp expression was shown to be limited to Th1 cells (Torigoe et al. 1997; Xu et al. 1998). Th1 cells derived from IL-1Rrp-knockout mice failed to activate NFκB in response to IL-18, and their NK cells showed decreased cytolytic activity and decreased IFN-γ produced in response to *P. acnes* (Hoshino et al. 1999). A second chain of the IL-18 receptor has been identified and designated as IL-1 receptor accessory protein-like (IL-1RAcPL). Both IL-1Rrp and AcPL are required for IL-18 mediated activation of NFκB and c-jun N-terminal kinase (Born et al. 1998; Debets et al. 2000). IL-1Rrp and IL-1RAcPL are expressed by Th1 cells and downregulated in Th2 cells (Debets et al. 2000). Thus, distinct tissue specificity of IL-1α/β and IL-18 receptors allows these cytokines to utilize a common signaling pathway in distinct Th cell subsets. Interestingly, the IRAK/NFκB signaling pathway is highly conserved throughout evolution and homologous pathways have been found in species as diverse as plants, *Drosophila*, and mammals (O'Neill and Greene 1998).

The importance of IL-18 in Th1 responses has been shown for a variety of pathogens. IL-18 knockout mice were more susceptible to bacillus Calmette-Guerin (BCG) (Takeda et al. 1998), *Leishmania major* and to septic arthritis induced by *Staphlococcus aureus* (Wei et al. 1999). Moreover anti-IL-18 treatment exacerbated *Salmonella* infection (Dybing et al. 1999; Mastroeni et al. 1999) and IL-18 protein treatment ameliorated *Yersinia* and *Vaccinia* infections (Bohn et al. 1998; Tanaka-Kataoka et al. 1999). Anti-IL-18R antibody was shown to reduce local joint inflammation induced with carrageenin and to reduce mortality associated with LPS-induced shock (Xu et al. 1998). When compared with IL-12 knockout mice in their responses to BCG, IL-18-deficient mice showed similarly reduced responses as measured by IFN-γ production in a recall assay. Production of IFN-γ was almost totally abrogated in mice lacking both IL-12 and IL-18, suggesting that IL-12 and IL-18 play distinct and non-redundant roles at least in the response to this pathogen in mice (Takeda et al. 1998). The molecular mechanism underlying the synergy between IL-18 and IL-12 may be explained in part by observations that IL-18 upregulated the IL-12R (Xu et al. 1998), and IL-12 upregulated the IL-18R (Ahn et al. 1997; Yoshimoto et al. 1998). However, more studies are required to fully understand the molecular nature of the synergy between IL-12 and IL-18 in inducing IFN-γ.

3 Restoration of IL-12 Signaling in Th2 Cells Does Not Lead to IFN-γ Production

The regulation of cytokine receptor expression in Th1 cell development was first observed with the IL-12R. As previously discussed, the IL-12Rβ2 chain was downregulated upon differentiation to the Th2 phenotype preventing IL-12 signaling in these cells (ROGGE et al. 1997; SZABO et al. 1997). This downregulation of the IL-12R was proposed to be a key checkpoint in preventing Th2 cells from responding to IL-12 and producing IFN-γ (SZABO et al. 1997). We tested this hypothesis by using retroviral infection to ectopically express the IL-12Rβ2 chain in both developing Th2 cells and a committed Th2 clone (HEATH et al. 2000). These cells constitutively expressed the IL-12Rβ1 chain and we showed that the ectopic expression of IL-12Rβ2 in Th2 cells was sufficient to restore STAT4 signaling and proliferation. However, the addition of IL-12 to IL-12Rβ2-expressing Th2 cells induced only a very low level of IFN-γ and did not decrease the amount of IL-4 produced (HEATH et al. 2000). This was found to be the case for 1-, 2- and 3-week polarized Th2 cells derived from a TCR transgenic mouse and for a Th2 clone. We also tested the possibility that Th2 cells expressing the IL-12R could be switched to a Th1 phenotype with prolonged culture in IL-12 plus anti-IL-4. We found, however, that after 1 week of culture in Th2 conditions, it was not possible to switch the phenotype of these cells and induce them to produce IFN-γ despite their forced expression of the IL-12R. These data suggested that IL-12 signaling alone is insufficient to induce the production of IFN-γ in Th2 cells, implying that other factors are required for the expression of this cytokine in these cells. Previous studies showed that developing Th2 cells in the presence of either IL-12 or IFN-γ was sufficient to abrogate the downregulation of IL-12Rβ2 normally seen upon culture in IL-4 (SZABO et al. 1997). These cells produced IL-4 and some IFN-γ upon restimulation, and IFN-γ production could be increased by restimulating the cells in the presence of IL-12. In the case of Th cells developed in IL-4 and IL-12, it was found that IFN-γ production upon restimulation with splenic APC, antigen and IL-12 could be augmented by the addition of anti-IL-10 and anti-IL-4 mAbs. It is not clear, however, whether IL-4 and IL-10 were downregulating APC-derived factors and/or receptors or transcription factors in Th cells required for IFN-γ production (SZABO et al. 1997). With respect to our own data, it is possible that there is a requirement for IL-18 signaling as well as IL-12 signaling in the induction of IFN-γ production by Th2 cells. As discussed earlier, the IL-18 receptor is downregulated on Th2 cells, and so the role of

IL-18 could not be assessed. Also relevant to this discussion is the recent identification of a novel transcription factor of the brachyury family, named T-bet, which is expressed in Th1 but not Th2 cells. Ectopic expression of T-bet in developing as well as committed Th2 cells resulted in both the induction of IFN-γ and the reduction of IL-4 and IL-5 production (SZABO et al. 2000).

4 Does STAT-6 Activation and GATA-3 Expression in Th1 Cells Restore Production of Th2-Specific Factors?

In order to gain further insight into the mechanisms of Th1 development, we analyzed the extent to which various Th2 signaling pathways could modify the Th1 phenotype. To this end we restored components of the IL-4 signaling pathway or forced expression of Th2 specific transcription factors in Th1 cells. IL-4 has been identified as the dominant cytokine in Th2 development, and this cytokine induces activation of STAT6 and phosphorylation of SHP (Src homology 2 domain-containing 5′ inositol phosphatase), IRS-1/2 and shc (RYAN et al. 1996; WANG et al. 1996; ZAMORANO and KEEGAN 1998). Both IL-4 and its induction of the STAT6 signaling pathway were shown to be important in Th2 development since both IL-4 and STAT6 knockout mice have impaired Th2 responses (KUHN 1991; QUELLE et al. 1995; KAPLAN et al. 1996; SHIMODA 1996). Since Th1 cells express the IL-4 receptor but fail to activate STAT6 when treated with IL-4 (KUBO et al. 1997; HUANG and PAUL 1998), we were interested to assess the effects of restoring STAT6 signaling in Th1 cells. In order to do this an inducible form of the transcription factor STAT6 was used. STAT6 was fused to the estrogen receptor (STAT6:ER) and this molecule was previously shown to be activated upon addition of the estrogen analogue 4-hydroxy-tamoxifen (4-HT) (KAMOGAWA et al. 1998). Expression of STAT6:ER in developing Th1 cells downregulated IFN-γ production and increased expression of IL-4 and IL-5 in a 4-HT dependent manner (KURATA et al. 1999). This was associated with upregulation of the Th2-specific transcription factors GATA-3 and c-maf and downregulation of the IL-12Rβ2 chain. We have previously shown that repeatedly stimulating Th1 cells in the presence of IL-12, or Th2 cells in the presence of IL-4, resulted in populations of T cells which were homogeneous in terms of their cytokine production and increasingly refractory to being switched to the opposite phenotype (MURPHY et al. 1996). Interestingly, as Th1 cells are repeatedly stimulated and polarized they became progressively resistant to induction of GATA-3, c-maf, and IL-4, and downregulation of IFN-γ, by activation of STAT6:ER (Fig. 3). A fully committed STAT6:ER-expressing Th1 clone could not be induced to produce IL-4 when treated for a prolonged period with 4-HT. In these experiments, a lack of IL-4 production correlated with the absence of GATA-3 and c-maf induction. These results are reminiscent of that seen when the IL-12 signaling pathway is restored in committed Th2 cells, suggesting that both Th1 and Th2 cells terminally differentiate in such a way as to become refractory to IL-4 and IL-12 cytokine signaling, respectively.

GATA-3 and c-maf are two transcription factors which have been identified over the last few years to be specifically expressed by Th2 cells and not Th1 cells (Ho et al. 1996; Zheng and Flavell 1997). It has been shown that ectopic expression of c-maf as a transgene in mice did not induce IL-4 in Th1 cells developed from such mice (Ho et al. 1998). GATA-3 was originally identified as a transcription factor that regulates TCR-α gene (Ho et al. 1991) and is essential for the development of the T cell lineage (Ting et al. 1996). Subsequently GATA-3 was shown to be Th2 cell specific (Zheng and Flavell 1997; Zhang et al. 1997; Lee et al. 1998), and able to strongly transactivate the IL-5 promoter but only weakly activate the IL-4 promoter (Lee et al. 1998; Ranganath et al. 1998; Zhang et al. 1998). We wanted to determine whether, unlike STAT6 activation, expression of GATA-3 in committed Th1 cells could induce Th2 cytokines and/or decrease expression of IFN-γ in developing as well as committed Th1 cells. As predicted, GATA-3 upregulated IL-4 and IL-5 expression in developing Th1 cells (Fig. 4). Somewhat more surprising was that IFN-γ expression was significantly reduced by ectopic expression of GATA-3 (Fig. 4) (Ouyang et al. 1998; Ferber et al. 1999). Thus GATA-3 not only upregulates production of Th2 cytokines (Zheng and Flavell 1997; Ouyang et al. 1998; Ferber et al. 1999), but also inhibits IFN-γ expression, and this has been shown in both IL-4 and STAT6 knockout mice (Ouyang et al. 1998, 2000; Ferber et al. 1999). Ouyang et al., showed that GATA-3 acted during the early stages of Th1 cell development to downregulate the IL-12Rβ2 thus making the cells refractory to IL-12 (Ouyang et al. 1998). This was proposed to be one mechanism by which GATA-3 could downregulate Th1 development. Of particular importance was our observation that expression of GATA-3, unlike activation of STAT6:ER, in a Th1 clone and 3x polarized Th1 cells, could induce IL-4 expression and diminish IFN-γ production (H.-J. Lee et al., submitted). This effect was more clearly seen when cyclic adenosine monophosphate (cAMP) was added to the cultures, with levels of IL-4 production approaching that of

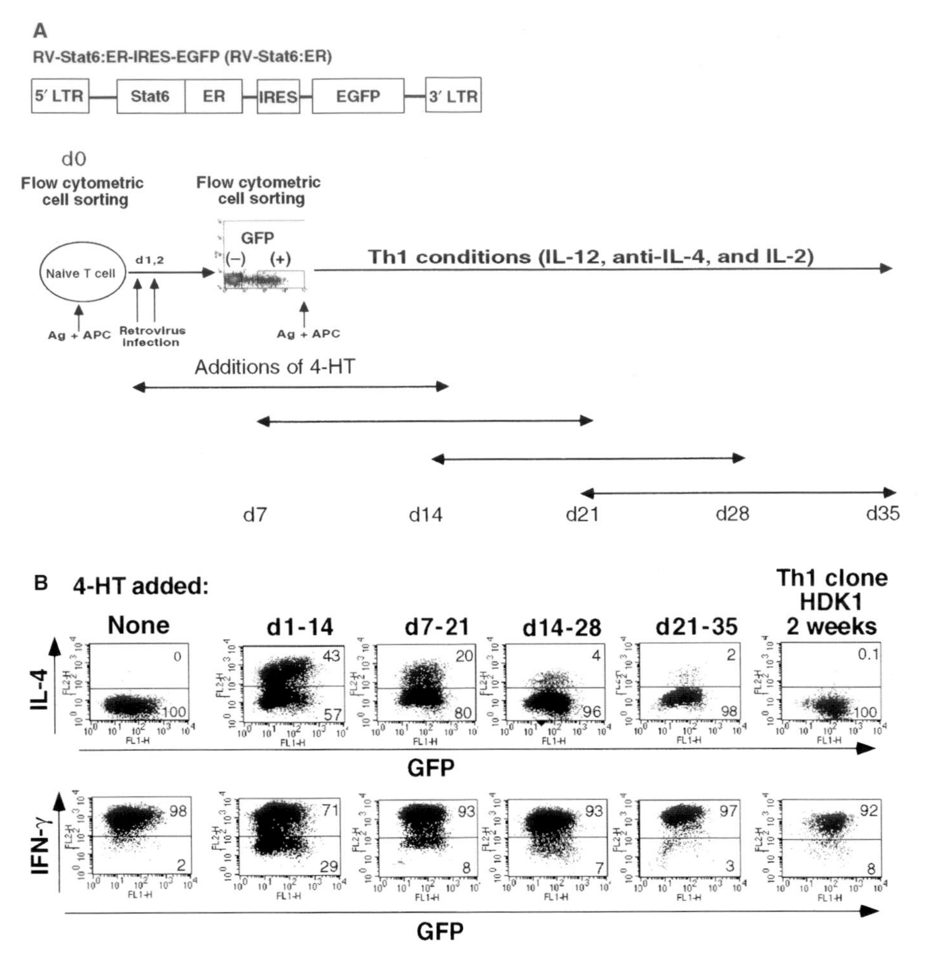
A
RV-Stat6:ER-IRES-EGFP (RV-Stat6:ER)
5′ LTR
Stat6
ER
IRES
EGFP
3′ LTR
d0
Flow cytometric cell sorting
Flow cytometric cell sorting
GFP
(−) (+)
Th1 conditions (IL-12, anti-IL-4, and IL-2)
Naive T cell
d1,2
Ag + APC
Retrovirus infection
Ag + APC
Additions of 4-HT
d7
d14
d21
d28
d35
B 4-HT added:
None
d1-14
d7-21
d14-28
d21-35
Th1 clone HDK1 2 weeks
IL-4
GFP
IFN-γ
GFP

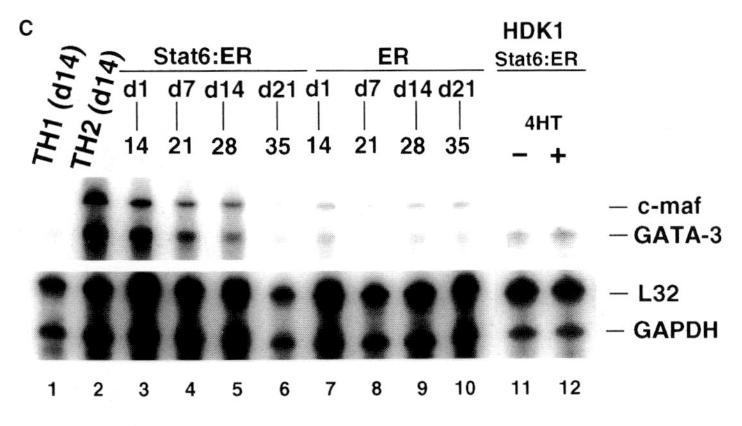
C
TH1 (d14)
TH2 (d14)
Stat6:ER
ER
HDK1 Stat6:ER
d1 d7 d14 d21 d1 d7 d14 d21
14 21 28 35 14 21 28 35
4HT
− +
c-maf
GATA-3
L32
GAPDH
1 2 3 4 5 6 7 8 9 10 11 12

Fig. 3. The induction of the Th2 phenotype by STAT6:ER is limited to the early stages of development. **A** Schematic of the experimental procedure. Naïve DO11.10 CD4$^+$ T cells were sorted and stimulated with splenic APC, antigen and IL-12, and anti-IL-4. On days 1 and 2 they were retrovirally transduced with retroviral (*RV*)-STAT6:ER linked via an internal ribosomal entry site (*IRES*) to enhanced fluorescent green protein (*EGFP*) allowing bicistronic expression of these two genes. At day 7 the EGFP$^+$ cells were sorted and taken through further rounds of stimulation. 4-HT was added to activate STAT6:ER for 2 week periods from days 1–14, 7–21, 14–28, and 21–35. Assays were also performed with the HDK1 Th1 clone which was also transduced with RV-STAT6:ER and cultured for 2 weeks with 4-HT. **B** Cells were assayed at the end of these time periods by intracellular cytokine staining for IL-4 and IFN-γ. **C** RNA from these different populations was prepared and used in an RNAse protection assay to measure levels of GATA-3 and c-maf RNA

Th2 clones (H.-J. Lee et al., submitted). Retroviral expression of GATA-3 was found to activate expression of the endogenous GATA-3 gene but not c-maf, suggesting that GATA-3 autoregulates its own expression (H.-J. Lee et al., submitted). Thus, GATA-3 appears to be a key regulator of the Th2 phenotype, possibly by playing a role in chromatin remodeling.

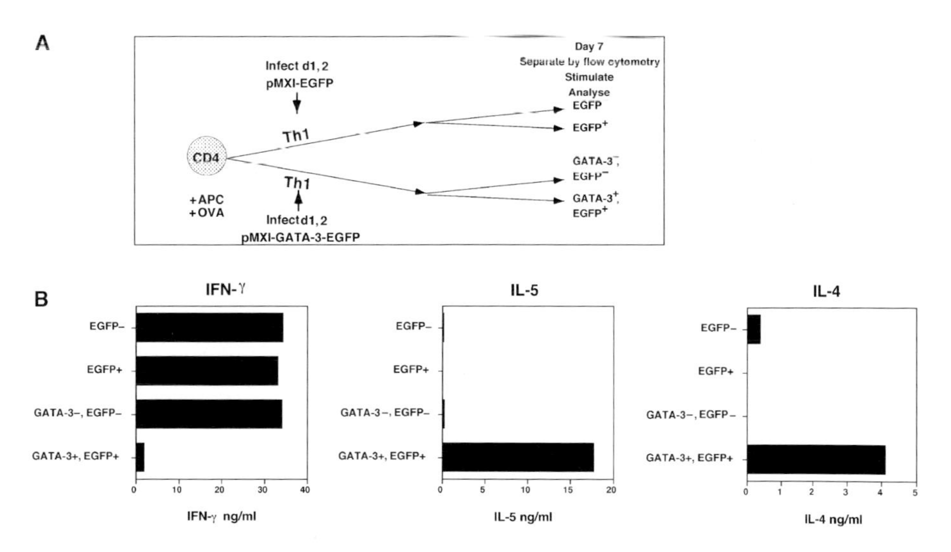

Fig. 4. GATA-3 inhibits IFN-γ and enhances IL-4 and IL-5 production in Th1 cells. **A** A scheme depicting the experimental protocol. Naïve DO11.10 CD4$^+$ T cells were sorted and stimulated with splenic APC, antigen and IL-12, and anti-IL-4. On days 1 and 2 they were retrovirally transduced with RV-GATA-3-EGFP or the control virus expressing EGFP alone. At day 7 the EGFP$^-$ and EGFP$^+$ cells were sorted to give four populations of cells, EGFP$^-$, EGFP$^+$, GATA-3-EGFP$^-$ or GATA-3EGFP$^+$. **B** These cells were restimulated with PMA and ionomycin for 48h, the supernatants were harvested, and IL-4, IL-5, and IFN-γ were measured by ELISA

5 Summary

We have described the work of our own and others characterizing some of the mechanisms of Th1 and Th2 development (Fig. 5). IL-12 is a key cytokine in driving Th1 cell development, and in certain mouse strains such as the BALB/c mouse, IL-1α and IL-18 act as cofactors in this process. IL-1α and IL-18, although belonging to the IL-1 family of cytokines, have very different actions on differentiated Th1 and Th2 cells. IL-1α can induce proliferation of Th2 cells, while IL-18 synergizes with IL-12 to induce maximal IFN-γ production by Th1 cells. IL-1α and IL-18 bind to distinct and differentially regulated receptors on Th2 versus Th1 cells, respectively, and induce IRAK phosphorylation and NFκB nuclear translocation. Dissection of the molecular mechanisms of Th1 and Th2 development led to the observations that ectopic expression of IL-12Rβ2 in Th2 cells, although it resulted in the restoration of IL-12 signaling and proliferation, did not allow IL-12 induced IFN-γ production nor downregulation of IL-4 production. The expression and activation of a conditionally active form of STAT6 in developing Th1 cells resulted in reduced IFN-γ and enhanced IL-4 production; however, the phenotype of committed Th1 cells could not

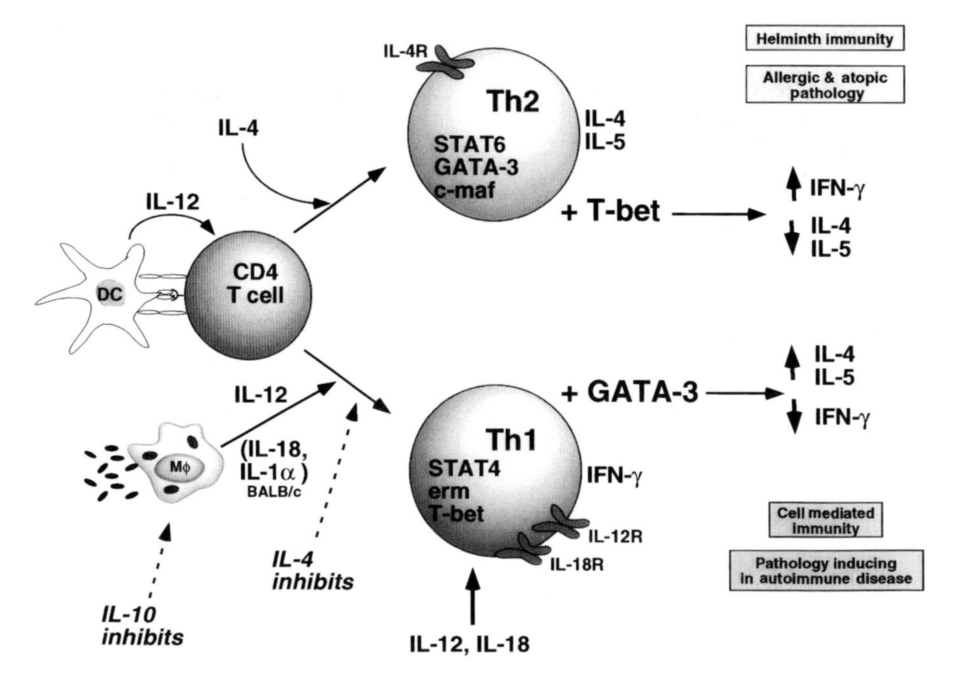

Fig. 5. Checkpoints and commitment factors for Th responses

be altered. In contrast ectopic expression of the Th2-specific transcription factor GATA-3 was able to upregulate IL-4 and IL-5 in both developing and committed Th1 cells, suggesting that this transcription factor may play a key role in Th2 development.

Acknowledgements. DNAX Research Institute is supported by Schering-Plough Corporation. The authors are grateful to Dr. Michael Tomlinson for proofreading the manuscript.

References

Abbas AK, Murphy KM, Sher A (1996) Functional diversity of helper T lymphocytes. Nature 383:787–793

Ahn HJ, Maruo S, Tomura M, Mu J, Hamaoka T, Nakanishi K, Clark S, Kurimoto M, Okamura H, Fujiwara H (1997) A mechanism underlying synergy between IL-12 and IFN-γ-inducing factor in enhanced production of IFN-γ. J Immunol 159:2125–2131

Bacon CM, Petricoin EF 3rd, Ortaldo JR, Rees RC, Larner AC, Johnston JA, O'Shea JJ (1995) Interleukin 12 induces tyrosine phosphorylation and activation of STAT4 in human lymphocytes. Proc Natl Acad Sci USA 92:7307–7311

Bazan JF, Timans JC, Kastelein RA (1996) A newly defined interleukin-1? Nature 379:591

Bohn E, Sing A, Zumbihl R, Bielfeldt C, Okamura H, Kurimoto M, Heesemann J, Autenrieth IB (1998) IL-18 (IFN-γ-inducing factor) regulates early cytokine production in, and promotes resolution of, bacterial infection in mice. J Immunol 160:299–307

Born TL, Thomassen E, Bird TA, Sims JE (1998) Cloning of a novel receptor subunit, AcPL, required for interleukin-18 signaling. J Biol Chem 273:29445–29450

Debets R, Timans JC, Churakowa T, Zurawski S, de Waal Malefyt R, Moore KW, Abrams JS, O'Garra A, Bazan JF, Kastelein RA (2000) IL-18 receptors, their role in ligand binding and function: anti-IL-1RAcPL antibody, a potent antagonist of IL-18. J Immunol 165:4950–4956

Dybing JK, Walters N, Pascual DW (1999) Role of endogenous interleukin-18 in resolving wild-type and attenuated *Salmonella typhimurium* infections. Infect Immun 67:6242–6248

Ferber IA, Lee HJ, Zonin F, Heath V, Mui A, Arai N, O'Garra A (1999) GATA-3 significantly downregulates IFN-γ production from developing Th1 cells in addition to inducing IL-4 and IL-5 levels. Clin Immunol 91:134–144

Gery I, Lepe-Zuniga JL (1983) Interleukin 1. Lymphokines 9:109–126

Greenbaum LA, Horowitz JB, Woods A, Pasqualini T, Reich E, Bottomly K (1988) Autocrine growth of $CD4^+$ T cells. Differential effects of IL-1 on helper and inflammatory T cells. J Immunol 140:1555

Greenfeder SA, Nunes P, Kwee L, Labow M, Chizzonite RA, Ju G (1995) Molecular cloning and characterization of a second subunit of the interleukin 1 receptor complex. J Biol Chem 270:13757–13765

Heath VL, Showe L, Crain C, Barrat FJ, Trinchieri G, O'Garra A (2000) Cutting Edge: Ectopic expression of the IL-12-Receptor-β2 in developing and committed Th2 cells does not affect the production of IL-4 or induce the production of IFN-γ. J Immunol 164:2861–2865

Hilkens CM, Messer G, Tesselaar K, van Rietschoten AG, Kapsenberg ML, Wierenga EA (1996) Lack of IL-12 signaling in human allergen-specific Th2 cells. J Immunol 157: 4316–4321

Ho IC, Lo D, Glimcher LH (1998) c-maf promotes T helper cell type 2 (Th2) and attenuates Th1 differentiation by both interleukin 4-dependent and -independent mechanisms. J Exp Med 188:1859–1866

Ho I-C, Hodge MR, Rooney JW, Glimcher LH (1996) The proto-oncogene c-maf is responsible for tissue-specific expression of interleukin-4. Cell 85:973–983

Ho IC, Vorhees P, Marin N, Oakley BK, Tsai SF, Orkin SH, Leiden JM (1991) Human GATA-3: a lineage-restricted transcription factor that regulates the expression of the T cell receptor alpha gene. Embo J 10:1187–1192

Hoshino K, Tsutsui H, Kawai T, Takeda K, Nakanishi K, Takeda Y, Akira S (1999) Cutting edge: generation of IL-18 receptor-deficient mice: evidence for IL-1 receptor-related protein as an essential IL-18 binding receptor. J Immunol 162:5041–5044

Hsieh C-S, Macatonia SE, Tripp CS, Wolf SF, O'Garra A, Murphy KM (1993) Development of Th1 CD4+ T cells through IL-12 Produced by Listeria-induced macrophages. Science 260:547–549

Huang H, Paul WE (1998) Impaired interleukin 4 signaling in T helper type 1 cells. J Exp Med 187:1305–1313

Hunter CA, Timans J, Pisacane P, Menon S, Cai G, Chizzonitte R, Bazan JF, Kastelein RA (1997) Comparison of the effects of interleukin-1α Interleukin-1β and Interferon-γ inducing factor on the production of interferon-γ by natural killer cells. Eur J Immunol 27:2787–2792

Jacobson NG, Szabo SJ, Weber-Nordt RM, Zhong Z, Schreiber RD, Darnell JEJ, Murphy KM (1995) Interleukin 12 signalling in T helper type 1 (Th1) cells involves tyrosine phosphorylation of signal transducer and activator of transcription (stat)3 and Stat4. J Exp Med 181:1755–1762

Kamogawa Y, Minasi L-aE, Carding SR, Bottomly K, Flavell RA (1993) The relationship of IL-4 and IFN-γ-producing T cells studied by lineage ablation of IL-4 producing cells. Cell 75:985–995

Kamogawa Y, Lee HJ, Johnston JA, McMahon M, O'Garra A, Arai N (1998) A conditionally active form of STAT6 can mimic certain effects of IL-4. J Immunol 161:1074–1077

Kaplan M, Schindler U, Smiley ST, Grusby MJ (1996) Stat6 is required for mediating responses to IL-4 and for the development of Th2 cells. Immunity 4:313–319

Kubo M, Ransom J, Webb D, Hashimoto Y, Tada T, Nakayama T (1997) T-cell subset-specific expression of the IL-4 gene is regulated by a silencer element and STAT6. Embo J 16:4007–4020

Kuhn R, Rajewsky K, Muller W (1991) Generation and analysis of interleukin-4 deficient mice. Science 254:707–710

Kurata H, Lee HJ, O'Garra A, Arai N (1999) Ectopic expression of activated Stat6 induces the expression of Th2-specific cytokines and transcription factors in developing Th1 cells. Immunity 11:677–688

Lee HJ, O'Garra A, Arai K, Arai N (1998) Characterization of *cis*-regulatory elements and nuclear factors conferring Th2-specific expression of the IL-5 gene: a role for a GATA-binding protein. J Immunol 160:2343–2352

Liblau R, Singer S, McDevitt H (1995) Th1 and Th2 CD4+ T cells in the pathogenesis of organ-specific autoimmune diseases. Immunol Today 16:34–38

Lichtman AH, Chin J, Schmidt JA, Abbas AK (1988) Role of interleukin 1 in the activation of T lymphocytes. Proc Natl Acad Sci USA 85:9699–9703

Magram J, Connaughton SE, Warrier RR, Carvajal DM, Wu C, Ferrante J, Stewart C, Sarmiento U, Faherty DA, Gately MK (1996) IL-12-deficient mice are defective in IFN-g production and type 1 cytokine responses. Immunity 4:471–481

Manetti R, Parronchi P, Guidizi MG, Piccinni M-P, Maggi E, Trinchieri G, Romagnani S (1993) Natural killer cell stimulatory factor (interleukin 12 [IL-12]) induces T helper type 1 (Th1)-specific immune responses and inhibits the development of IL-4-producing cells. J Exp Med 177:1199–1204

Mastroeni P, Clare S, Khan S, Harrison JA, Hormaeche CE, Okamura H, Kurimoto M, Dougan G (1999) Interleukin 18 contributes to host resistance and gamma interferon production in mice infected with virulent *Salmonella typhimurium*. Infect Immun 67: 478–483

McArthur JG, Raulet DH (1993) CD28-induced costimulation of T helper type 2 cells mediated by induction of responsiveness to interleukin 4. J Exp Med 178:1645–1653

Mosmann TR, Cherwinski H, Bond MW, Giedlin MA, Coffman RL (1986) Two types of murine helper T cell clone. I. Definition according to profiles of lymphokine activities and secreted proteins. J Immunol 136:2348–2357

Murphy E, Shibuya K, Hosken N, Openshaw P, Maino V, Davis K, Murphy K, O'Garra A (1996) Reversibility of T Helper 1 and T Helper 2 Cytokine Profiles is lost after long-term stimulation. J Exp Med 183:901–913

O'Garra A (1998) Cytokines induce the development of functionally heterogeneous T helper cell susbsets. Immunity 8:275–283

Okamura H, Tsutsul H, Komatsu T, Yutsudo M, Hakura A, Tanimoto T, Torigoe K, Okura T, Mukuda Y, Hattori K, Akita K, Namba M, Tanabe K, Konishi K, Fukuda S, Kurimoto M (1995) Cloning of a new cytokine that induces IFN-γ production by T cells. Nature 378:88–91

O'Neill LA, Greene C (1998) Signal transduction pathways activated by the IL-1 receptor family: ancient signaling machinery in mammals insects and plants. J Leukoc Biol 63: 650–657

Ouyang W, Ranganath SH, Weindel K, Bhattacharya D, Murphy TL, Sha WC, Murphy KM (1998) Inhibition of Th1 development mediated by GATA-3 through an IL-4-independent mechanism. Immunity 9:745–755

Ouyang W, Lohning M, Gao Z, Assenmacher M, Ranganath S, Radbruch A, Murphy KM (2000) Stat6-independent GATA-3 autoactivation directs IL-4-independent Th2 development and commitment. Immunity 12:27–37

Parnet P, Garka KE, Bonnert TP, Dower SK, Sims JE (1996) IL-1Rrp is a novel receptor-like molecule similar to the type I interleukin-1 receptor and its homologues T1/ST2 and IL-1R AcP. J Biol Chem 271:3967–3970

Powrie F, Coffman RL (1993) Cytokine regulation of T-cell function: potential for therapeutic intervention. Immunol Today 14:270–274

Presky DH, Yang H, Minetti LJ, Chua AO, Nabavi N, Wu CY, Gately MK, Gubler U (1996) A functional interleukin 12 receptor complex is composed of 2 beta-type cytokine receptor subunits. Proc Natl Acad Sci USA 93:14002

Quelle FW, Shimoda K, Thierfelder W, Fischer C, Kim A, Ruben SM, Cleveland JL, Pierce JH, Keegan AD, Nelms K, Paul WE, Ihle JN (1995) Cloning of murine Stat6 and human Stat6, Stat proteins that are tyrosine phosphorylated in responses to IL-4 and IL-3 but are not required for mitogenesis. Mol Cell Biol 15:3336–3343

Ranganath S, Ouyang W, Bhattarcharya D, Sha WC, Grupe A, Peltz G, Murphy KM (1998) GATA-3-dependent enhancer activity in IL-4 gene regulation. J Immunol 161:3822–3826

Robinson D, Shibuya K, Mui A, Zonin F, Murphy E, Sana T, Hartley SB, Menon S, Kastelein R, Bazan F, O'Garra A (1997) IGIF does not drive Th1 development, but synergizes with IL-12 for interferon-γ production, and activates IRAK and NF-κB. Immunity 7:571–581

Rocken M, Muller KM, Saurat J-H, Muller I, Louis JA, Cerottini J-C, Hauser C (1992) Central role for TCR/CD3 ligation in the differentiation of CD4+ T cells toward a Th1 or Th2 functional phenotype. J Immunol 148:47–54

Rogge L, Barberis-Maino L, Biffi M, Passini N, Presky DH, Gubler U, Sinigaglia F (1997) Selective expression of an interleukin-12 receptor component by human T helper 1 cells. J Exp Med 185:825–831

Romagnani S (1991) Human T_H1 and T_H2 subsets: doubt no more. Immunol Today 12: 256–257
Romagnani S (1994) Lymphokine production by human T cells in disease states. Ann Rev Immunol 12:227–257
Ryan JJ, McReynolds LJ, Keegan A, Wang L-H, Garfein E, Rothman P, Nelms K, Paul WE (1996) Growth and gene expression are predominantly controlled by distinct regions of the human IL-4 receptor. Immunity 4:123–132
Seder RA, Gazzinelli R, Sher A, Paul WE (1993) Interleukin 12 acts directly on $CD4^+$ T cells to enhance priming for interferon γ production and diminishes interleukin 4 inhibition of such priming. Proc Natl Acad Sci USA 90:10188–10192
Sher A, Coffman RL (1992) Regulation of immunity to parasites by T cells and T cell-derived cytokines. Ann Rev Immunol 10:385–409
Shibuya K, Robinson D, Zonin F, Hartley SB, Macatonia SE, Somoza C, Hunter CA, Murphy KM, O'Garra A (1998) IL-1 alpha and TNF-alpha are required for IL-12-induced development of Th1 cells producing high levels of IFN-gamma in BALB/c but not C57BL/6 mice. J Immunol 160:1708–1716
Shimoda K, van Deursen J, Sangster MY, Sarawar SR, Carson RT, Tripp RA, Chu C, Quelle FW, Nosaka T, Vignali DA Doherty PC, Grosveld G, Paul WE, Ihle JN (1996) Lack of IL-4-induced Th2 response and IgE class switching in mice with disrupted Stat6 gene. Nature 380:630–633
Sims JE, March CJ, Cosman D, Widmer MB, MacDonald HR, McMahan CJ, Grubin CE, Wignall JM, Jackson JL, Call SM, et al. (1988) cDNA expression cloning of the IL-1 receptor a member of the immunoglobulin superfamily. Science 241:585–589
Szabo S, Dighe AS, Gubler U, Murphy KM (1997) Regulation of the interleukin (IL)-12β2 Subunit expression in developing T helper 1 (Th1) and Th2 Cells. J Exp Med 185:817–824
Szabo SJ, Jacobson NG, Dighe AS, Gubler U, Murphy KM (1995) Developmental Commitment to the Th2 Lineage by Exticntion of IL-12 Signaling. Immunity 2:665–675
Szabo SJ, Kim ST, Costa GL, Zhang X, Fathman CG, Glimcher LH (2000) A novel transcription factor, T-bet, directs Th1 lineage commitment. Cell 655–669
Takeda K, Tsutsui H, Yoshimoto T, Adachi O, Yoshida N, Kishimoto T, Okamura H, Nakanishi K, Akira S (1998) Defective NK cell activity and Th1 response in IL-18-deficient mice. Immunity 8:383–390
Tanaka-Kataoka M, Kunikata T, Takayama S, Iwaki K, Ohashi K, Ikeda M, Kurimoto M (1999) In vivo antiviral effect of interleukin 18 in a mouse model of vaccinia virus infection. Cytokine 11:593–599
Thierfelder WE, van Deursen JM, Yamamoto K, Tripp RA, Sarawar SR, Carson RT, Sangster MY, Vignali DA, Doherty PC, Grosveld GC, Ihle JN (1996) Requirement for Stat4 in interleukin-12-mdeiated responses of natural killer and T cells. Nature 382:171–174
Ting CN, Olson MC, Barton KP, Leiden JM (1996) Transcription factor GATA-3 is required for development of the T-cell lineage. Nature 384:474–478
Torigoe K, Ushio S, Okura T, Kobayashi S, Taniai M, Kunikata T, Murakami T, Sanou O, Kojima H, Fujii M, Ohta T, Ikeda M, Ikegami H, Kurimoto M (1997) Purification and characterization of the human interleukin-18 receptor. J Biol Chem 272:25737–25742
Trinchieri G (1995) Interleukin-12: a proinflammatory cytokine with immunoregulatory functions that bridge innate resistance and antigen-specific adaptive immunity. Ann Rev Immunol 13:251–276
Wang HY, Paul WE, Keegan AD (1996) IL-4 function can be transferred to the IL-2 receptor by tyrosine containing sequences found in the IL-4 receptor α chain. Immunity 4:113–121
Wei XQ, Leung BP, Niedbala W, Piedrafita D, Feng GJ, Sweet M, Dobbie L, Smith AJ, Liew FY (1999) Altered immune responses and susceptibility to Leishmania major and *Staphylococcus aureus* infection in IL-18-deficient mice. J Immunol 163:2821–2828

Wu C, Ferrante J, Gately MK, Magram J (1997a) Characterization of IL-12 receptor beta1 chain (IL-12Rbeta1)-deficient mice: IL-12Rbeta1 is an essential component of the functional mouse IL- 12 receptor. J Immunol 159:1658–1665
Wu C, Warrier RR, Wang X, Presky DH, Gately MK (1997b) Regulation of interleukin-12 receptor beta1 chain expression and interleukin-12 binding by human peripheral blood mononuclear cells. Eur J Immunol 27:147–154
Xu D, Chan WL, Leung BP, Hunter D, Schulz K, Carter RW, McInnes IB, Robinson JH, Liew FY (1998) Selective expression and functions of interleukin 18 receptor on T helper (Th) type 1 but not Th2 cells. J Exp Med 188:1485–1492
Yoshimoto T, Takeda K, Tanaka T, Ohkusu K, Kashiwamura S, Okamura H, Akira S, Nakanishi K (1998) IL-12 up-regulates IL-18 receptor expression on T cells, Th1 cells, and B cells: synergism with IL-18 for IFN-γ production. J Immunol 161:3400–3407
Zamorano J, Keegan AD (1998) Regulation of apoptosis by tyrosine-containing domains of IL-4R alpha: Y497 and Y713, but not the STAT6-docking tyrosines, signal protection from apoptosis. J Immunol 161:859–867
Zhang DH, Yang L, Ray A (1998) Differential responsiveness of the IL-5 and IL-4 genes to transcription factor GATA-3. J Immunol 161:3817–3821
Zhang DH, Cohn L, Ray P, Bottomly K, Ray A (1997) Transcription factor GATA-3 is differentially expressed in murine Th1 and Th2 cells and controls Th2-specific expression of the interleukin-5 gene. J Biol Chem 272:21597–21603
Zheng W, Flavell RA (1997) The transcription factor GATA-3 is necessary and sufficient for Th2 cytokine gene expression in CD4 T cells. Cell 89:587–596

The Role of Complement in Innate and Adaptive Immunity

J.E. Volanakis

1 Introduction

The complement system is an ancient element of innate immunity present at progressively more evolved forms in the blood or lymph of all vertebrates studied and in at least some invertebrate species (Nonaka 1998). It provides one of the best examples of an effective first line of host defense system. Complement can recognize conserved repetitive structural elements important for the survival of pathogens and therefore shared by large groups of them. Recognition leads to the activation of the system and elimination of the pathogens either by direct killing or by opsonophagocytosis and intracellular killing. In parallel, complement also provides signals to adaptive immunity instructing it to respond vigorously against pathogens (Carroll 1998; Fearon 1998). Furthermore, through its ability

Division of Clinical Immunology and Rheumatology, Department of Medicine, University of Alabama at Birmingham, Alabama, USA
Present address: John E. Volanakis, Biomedical Sciences Research Center "A. Fleming", P.O. Box 74145, 16602 Varkiza, Greece

to discriminate between free and antigen-bound immunoglobulins, complement serves as an effector arm of adaptive humoral immunity leading to elimination of pathogens recognized by specific antibodies. Finally, complement appears to have a special relationship with natural antibodies depending on them for recognition of important pathogens and in turn providing signals for clonal selection and expansion of $CD5^+$ B-1 cells that produce natural antibodies (CARROLL and PRODEUS 1998).

In man, complement comprises about 35 plasma or membrane-associated proteins that participate in the activation or regulation of the system or serve as receptors for biologically-active products of complement activation. Proteins participating in complement activation act as recognition molecules, enzymes, enzyme cofactors, or precursors of active fragments. Regulatory proteins control the rate and extent of activation or inactivate biologically active products. Additionally, cell-membrane-associated regulatory proteins protect host cells from complement-mediated damage, constituting an effective self-recognition system (LISZEWSKI et al. 1996).

Activation of complement is necessary for expression of biologic activity. Activation is initiated by a wide variety of pathogens recognized either directly or indirectly by complement proteins and proceeds through pathways that culminate in the assembly of proteolytic enzymes termed convertases, which produce biologically active fragments (VOLANAKIS 1990, 1998). There are three activation pathways, termed classical, lectin, and alternative. Each pathway is activated by a range of pathogens and utilizes different enzymes and cofactors to form the convertases. Nevertheless, all convertases have the same substrate specificity for complement proteins C3 and C5, and all generate the same biologic activities from these proteins.

2 Role of Complement in Innate Immunity

Considering the presence of rudimentary complement systems in invertebrates (SMITH et al. 1996) and lower vertebrates, such as cyclostomes (ISHIGURO et al. 1992; NONAKA and TAKAHASHI 1992), which lack an adaptive immune system, it becomes obvious that complement has its roots in innate immunity. Indeed, its ability to recognize a wide range of pathogens and to eliminate them makes complement a major innate defense mechanism. The importance of complement in defense against pathogenic bacteria is indicated by the serious infectious complications associated with genetic deficiency of key complement proteins (ROSS and DENSEN 1984).

2.1 Pathogen Recognition

Two proteins, C1q and mannose-binding lectin (MBL), recognize pathogens in the classical and lectin pathway of complement activation, respectively. They belong to the "collectin" protein family (SASTRY and EZEKOWITZ 1993) and have very similar tertiary structures, although they don't appear to belong to the same gene family. Both proteins are oligomers of subunits formed by three polypeptide chains. C1q is a hexamer, while MBL ranges from dimer to hexamer. The three polypeptide chains forming the subunits of MBL are identical, while those of C1q are different, although structurally homologous. Each subunit has a trimeric globular head and a triple helical collagen-like "stalk". Globular heads contain recognition sites, while collagen stalks transmit activation signals to associated zymogens of serine proteases or to cellular receptors.

C1q recognizes a variety of pathogens, including gram-negative bacteria (LOOS et al. 1978; CLAS and LOOS 1981), mycoplasmas (BREDT et al. 1977), and retroviruses (COOPER et al. 1976; DIERICH et al. 1993). Recognition of gram-negative bacteria depends mainly on reactivity of C1q with the lipid A region of lipopolysaccharide (COOPER and MORRISON 1978) and/or with certain outer-membrane proteins (GALDIERO et al. 1984; MERINO et al. 1998), while HIV-1 binds C1q through the transmembrane protein gp41 (MARSCHANG et al. 1997). The C1q bacterial recognition repertoire expands significantly through its binding specificity for ligand-bound C-reactive protein (CRP). CRP is an acute-phase host-defense protein with binding specificity for phosphocholine residues (VOLANAKIS 1982). CRP can bind to *Pneumococcus pneumoniae* and to a variety of additional bacteria and other pathogens (SZALAI et al. 1999), which therefore also become targets for C1q. Fc structures of antigen-complexed IgG and IgM are also major targets of C1q, as discussed below.

MBL expresses lectin-like specificity for terminal mannose or *N*-acetylglucosamine residues. It has been shown to bind to the cell surface or purified fractions from cell surfaces of a wide range of pathogens. These include strains of gram-negative and gram-positive bacteria, yeasts, viruses, mycobacteria, and protozoa (EPSTEIN et al. 1996; TURNER 1996).

Binding of C1q or MBL to pathogens leads to the activation of associated serine proteases, C1r and C1s for C1q and MASP1 and/or MASP2 for MBL, which in turn act on C4 and C2 to form a C3/C5-convertase. In addition, both C1q and MBL act as opsonins, enhancing phagocytosis of their targets through binding to the same receptor, C1qRp, on phagocytic cells (TENNER et al. 1995; NEPOMUCENCO et al. 1997).

Bacteria and other pathogens also activate directly the alternative complement pathway, but through a different mechanism, which depends on the covalent binding of an activation fragment of C3, C3b, to the cell surface. C3b cannot discriminate between self and foreign pathogens, binding equally well to both. Recognition of pathogens is accomplished at the subsequent step through differential affinity of the regulatory protein factor H for surface bound C3b (Fearon and Austen 1977). On pathogen surfaces, C3b has diminished affinity for factor H and therefore can react with factor B and form an effective complement convertase. On host cells, C3b preferentially binds factor H, which precludes assembly of the convertase.

2.2 Acute Inflammation, Opsonophagocytosis, and Direct Killing of Pathogens

The acute inflammatory process is the principal means of host defense against infection by pyogenic bacteria. Complement can elicit all aspects of acute inflammation through its activation products C3a, C5a, and C5b-9. The first two are small peptides produced during activation of C3 and C5, respectively and are referred to as anaphylatoxins. C3a and C5a react with their respective receptors, C3aR (Ames et al. 1996) and C5aR (CD88) (Gerard and Gerard 1991), on mast cells and basophils to cause the release of vasoactive amines, such as histamine and serotonin, as well as leukotrienes (Ember et al. 1998). These mediators elicit vascular changes characteristic of inflammation such as, vasodilatation and increased permeability. Both anaphylatoxins also have multiple effects on myeloid cells. C5a is chemotactic for granulocytes and monocytes and causes upregulation of β_2 integrins thus promoting leukocyte adhesion (Kishimoto et al. 1989). In addition, C5a induces degranulation of granulocytes and also activates the reduced nicotinamide-adenine dinucleotide phosphate (NADPH)-oxidase pathway, leading to an oxidative burst (Goetzl and Austen 1974). Effects of C3a seem to be more selective for eosinophils and include chemotaxis, oxidative burst, and degranulation of these cells (Elsner et al. 1994; Daffern et al. 1995).

Phagocytosis and intracellular killing is the most important defense mechanism against bacteria and therefore enhancement of phagocytosis is arguably the most important contribution of complement to host defense. Although, as mentioned above, C1q and MBL enhance phagocytosis, the major complement opsonins are activation products of C3. Opsonic activity of C3 depends critically on the ability of its major activation fragment, C3b, to form ester bonds with free hydroxyl groups on bacterial surfaces through a transesterification reaction involving a metastable thioester bond. C3b and its further activation product, iC3b, are recognized by their respective receptors, CR1 (CD35) and CR3 (CD11b/CD18), on phagocytic cells and greatly enhance phagocytosis. Expression of both receptors is significantly upregulated by C5a and other chemotactic factors. It is generally believed that engagement of CR1 and/or CR3 does not deliver phagocytic signals unless an additional phagocytic receptor is also engaged. A classic example of such a "second" signal is provided by IgG engaging FcR (Bianco et al. 1975; Goldstein et al. 1976), but can also be provided by signaling through other receptors (Czop and Austen 1980).

The assembly of C5b-9, a large protein–protein complex of five proteins, C5b, C6, C7, C8, and C9, is initiated by C5b, a large activation fragment of C5 (Plumb and Sodetz 1998). During assembly of the complex, which is termed "membrane attack complex", hydrophobic regions of the participating proteins become exposed on its surface. Assembly on the surface of biologic membranes favors interactions of these hydrophobic regions with the fatty acid chains of phospholipids. The complex becomes gradually inserted into the lipid bilayer and eventually forms a transmembrane channel, which can elicit cellular functions (Niculescu et al. 1997) or lead to killing of susceptible cells. Among bacteria, gram-negative strains are particularly susceptible to direct killing by complement. In terms of host defense, this mechanism is of greater importance for defense against neisserial infections as indicated by the recurrent bouts of meningococcal meningitis and gonococcemia experienced by individuals with genetic deficiencies of proteins participating in the assembly of the membrane attack complex (Ross and Densen 1984). The C5b-9 complex also causes drastic changes of the microvasculature by directly activating endothelial cells leading to formation of endothelial gaps, the release of procoagulant activity, and the adhesion of leukocytes (Hamilton et al. 1990; Saadi and Platt 1995).

3 Interactions of Complement with Adaptive Immunity

Complement owes its name to its role as an effector arm of antibodies and the original impression of the people who discovered it that "complementing" the bacteriolytic action of antibodies was its only function. Much later it was realized that complement not only has multiple antibody-independent functions but also plays an important role in the regulation of humoral immune responses. Although the interactions between complement and adaptive immunity are not yet understood fully, considerable details on their nature and the molecular mechanisms involved have emerged recently.

3.1 Recognition of Immunoglobulin Complexes

IgG and IgM antibodies in complex with their respective antigens are the most extensively studied and perhaps the most effective activators of the classical pathway of complement. The binding sites on the globular heads of C1q recognize the Cγ2 domains of human IgG1, 2, or 3 (Duncan and Winter 1988) or the Cμ3 domain of IgM (Arya et al. 1994) antibodies. Human IgG4 antibodies, like mouse IgG1, are not recognized by C1q. Engagement of two or more of its globular heads induces a conformational change in C1q, which initiates a series of enzymatic reactions culminating in the assembly of a C3/C5 convertase. The end result is the decoration of the pathogen with large numbers of covalently linked C3b and iC3b, the major complement opsonins, and in the case of susceptible bacterial strains their direct killing by the C5b-9 complex.

In terms of early defense against pathogens, activation of the classical pathway by natural antibodies plays an important role. Natural antibodies are mainly of the IgM class, recognize pathogens, even if the host has never been exposed to them and are generally considered part of the innate rather than the adaptive immune system (Pereira et al. 1986). They are produced by B-1 cells, a distinct subset of self-replenishing B cells (Baumgarth et al. 1999) characterized by the expression of CD5 and CD11b. B-1 cells produce the major response to microorganismal coat antigens, such as lipopolysaccharide, α1–3 dextran, phosphorylcholine, and undefined determinants of *E. coli* and *Salmonella* (Kantor and Herzenberg 1993). In contrast to conventional B cells, exposure of B-1 cells to antigen does not stimulate somatic mutation, affinity maturation, and class switching. Characteristically, repeated exposure of B-1 cells to the same antigen elicits similar responses each time, and the responses are elicited within 48h. Mice made deficient in CD19 by gene targeting (Engel et al. 1995; Rickert et al. 1995) or by prenatal injection of anti-CD19 monoclonal antibody (Krop et al. 1996) have a striking reduction in B-1 cells, while conventional B cells appear normal. Reduced numbers of B-1 cells are also observed in mice deficient in the complement receptor CD21, which on B cells forms a signal-transducing complex with CD19, CD81, and Leu 13 (Ahearn et al. 1996). Thus, complement-activation-produced ligands may be essential for self-renewal and maintenance of B-1 cells.

3.2 Regulation of Immune Responses

Initial evidence indicating that complement plays a role in regulation of antibody responses was provided by PEPYS (1972) who reported that mice transiently depleted of complement by treatment with cobra venom factor had diminished antibody responses to T-dependent antigens.

A mechanism for this effect was suggested by the subsequent finding that complement was necessary for localization of immunoglobulin aggregates in germinal centers (PAPAMICHAIL et al. 1975). Transient depletion of complement was also found to abrogate development of memory B cells (KLAUS and HUMPHREY et al. 1977). Subsequent studies in humans (JACSKON 1979) guinea pigs (OCHS et al. 1983; BOETTGER et al. 1986) and dogs (O'NEIL et al. 1988) genetically deficient in complement proteins C2, C3, or C4 confirmed that the presence of an intact complement system is necessary for a normal antibody response to T-dependent antigens. Specifically, complement was shown to be important for responses to limited, but not high, concentrations of antigen. The role of complement in induction of humoral immune responses to small amounts of antigen was confirmed more recently by using mice bearing targeted disruptions of the genes encoding complement proteins C3 or C4 (FISCHER et al. 1996). The immune defect of these animals is characterized by decreased primary and secondary antibody responses, failure in isotype switching and a reduced number of germinal centers within splenic follicles. An understanding of the mechanism for the complement immunoregulatory effects has emerged from studies on the structural biology of complement receptors and the availability of gene-targeted mice deficient in these receptors.

3.2.1 Role of Complement Receptor Type 2 (CR2, CD21)

It is now well established that the main mechanism for complement-dependent regulation of B cell responses involves interactions between antigen-linked C3 activation fragments iC3b, C3dg, and C3d and complement receptor type 2 (CR2, CD21). C3dg and C3d are the final activation fragments of antigen-bound C3b and are produced by the action of the serine protease factor I in the presence of one of its cofactors. Their immediate precursor is iC3b, the main complement opsonin. CR2 is a 140-kDa transmembrane glycoprotein consisting of 15 or 16 tandemly repeated short consensus repeats (SCR), also termed complement control protein (CCP) modules, followed by a transmembrane segment and a short 34 amino acid intracytoplasmic tail (MOORE et al. 1987; WEIS et al. 1988). Human CR2 is encoded by a single gene and is expressed on B cells, follicular dendritic cells (FDC) and a subset of T cells. Complement receptor type 1 (CR1; CD35), a 190-kDa glycoprotein, has a structure very similar to CR2. It consists of 23–44 tandemly repeated CCP modules, followed by transmembrane and short intracytoplasmic domains (KLICKSTEIN et al. 1987). Human CR1 is encoded by a single gene, has binding specificity for C3b and C4b and a wider cellular distribution than CR2, being expressed on erythrocytes, myelomonocytic cells, B and some T cells, FDCs, and glomerular podocytes. In mice, CR1 and CR2 are differentially spliced products of a single gene (KURTZ et al. 1990), consisting of 21 and 15 CCP modules, respectively. Their binding specificities and cellular distributions are similar to those of their human homologs, except that CR1 is not expressed on mouse erythrocytes. On B cells CR2 forms a signal transducing complex with CD19, TAPA-1 (CD81), and Leu 13 (MATSUMOTO et al. 1991; BRADBURY et al. 1992). Also present on B cells is a CR1-CR2 complex, which has no apparent signal transduction properties (TUVESON et al. 1991).

A possible role for CR2 in regulation of B cell responses was suggested by experiments demonstrating that a monoclonal anti-mouse CR1/CR2 antibody, which blocked C3d binding to CR2, suppressed immune responses to suboptimal doses of T-dependent antigens (HEYMAN et al. 1990; THYPHRONITIS et al. 1991). Similar suppression of T-dependent B cell responses were observed by using a soluble form of CR2, which competed with cellular CR2 for ligand binding (HEBELL et al. 1991). In contrast to the inhibitory effects of CR2 blocking agents, 1,000- to 10,000-fold higher antibody responses were elicited by hen egg lysozyme fused to two or three molecules of C3d, respectively, than by lysozyme alone. The enhancing effect of the antigen-linked C3d was attributed to crosslinking the B cell antigen receptor complex to the CR2/CD19 complex (DEMPSEY et al. 1996). Evidence supporting this interpretation has been provided by in vitro experiments. Co-ligation of CR2 with membrane IgM on human lymphoblastoid cells resulted in synergistic enhancement of the release of intracellular calcium (CARTER et al. 1988). Similarly, coligation of CD19 with surface IgM on blood B cells lowered the threshold of B cell activation and increased the magnitude of the B cell proliferative response to optimal levels of IgM stimulation (CARTER and FEARON 1992). Downstream signaling events involved in the synergistic enhancement of antigen receptor-CR2/CD19 complex coligation include activation of CD19-associated tyrosine kinases leading to CD19 phosphorylation and the recruitment of Vav and PI-3 kinase and activation of PLC, which results in the release of intracellular calcium (review in TEDDER et al. 1997). Although CD19 may have independent functions, coligation with the antigen receptor complex is required for a positive effect on c-Jun N-terminal kinase or Bcl-2 expression and on ERK2 activation, DNA synthesis, and antibody secretion by mature B cells (Li and Carter 1998; Roberts and Snow 1999). The question of whether or not CD19 by itself has ligand-binding properties remains unanswered, since no ligand has been identified. Therefore, at present antigen-bound C3d is the only known physiologic ligand able to crosslink the antigen receptor and the CD19-CR2 complexes.

Additional insights into the role of signaling through the CR2/CD19 complex in B cell responses were provided by gene targeting of the CR2 and CD19 genes. CR2-deficient mice, which are also deficient in CR1, exhibit severe defects in primary and secondary antibody responses to T-dependent antigens (Ahearn et al. 1996; Molina et al. 1996). IgM antibody levels are moderately decreased, while IgG responses, particularly IgG1 and IgG3, are significantly reduced. One of the two strains of CR2-deficient mice has a substantial (40%) reduction in peritoneal B-1 cells and severe reduction in the number, size, and persistence of germinal centers developing in splenic follicles during secondary antigen challenge. CD19 gene-targeted mice have a similar, though more severe, phenotype, again suggesting CR2-independent CD19 functions (Engel et al. 1995; Rickert et al. 1995). A prominent feature of CD19-deficient mice is the pronounced reduction of peritoneal B-1 cells, which in normal mice exhibit the highest levels of CD19. In addition, CD19-deficient mice have reduced levels of serum immunoglobulins; their responses to T-dependent antigens are profoundly impaired, while responses to T-independent antigens are variable (Sato et al. 1995). Reduced humoral responses correlate with a scarcity of secondary follicles associated with impaired formation of germinal centers both before and after antigenic challenge.

Since CR2 is expressed on both B cells and FDCs, the question is raised as to which of the two cell types is responsible for the immune deficit of CR2-deficient mice. The role of B cell CR2 was indicated from reconstitution experiments using bone marrow from $CR2^{+/+}$ MHC-matched donors. Reconstituted mice had CR2 on B cells but not on FDCs and their antibody responses to a T-dependent antigen appeared normal (Ahearn et al. 1996). Further confirmation of the importance for B cell CR2 in immune responses was provided by mice produced by RAG-2-deficient blastocyst complementation (Croix et al. 1996). These animals specifically lack CR1 and CR2 on B cells but express both receptors on FDCs. They have a profound deficit in antibody responses to T-dependent antigens. Recent data have indicated that expression of CR2 provides not only an enhancing signal for lowering the threshold of B cell activation, but is also necessary for B cell survival within germinal centers and their differentiation along the memory pathway (FISCHER et al. 1998). A role for CR1 and CR2 on FDC has been suggested by the failure of CR2-deficient mice to develop memory B cells and the scarcity of immune complexes from germinal centers of complement-deficient animals. Direct demonstration of the need for CR1/CR2 expression on FDCs for an intact immune response has been provided by bone marrow reconstitution experiments using lethally irradiated $CR2^{-/-}$ and $CR2^{+/+}$ mice, which allowed the production of mice expressing the receptors on either B cells or FDCs, in both or neither cell type (FANG et al. 1998). These experiments demonstrated that expression of the receptors on FDCs is important primarily during the latter stages of the primary immune response and in the generation of a strong secondary response. The immune deficit of mice lacking the receptors in their FDCs correlated with impaired trapping of immune complexes in the splenic follicles, although germinal center development appeared normal.

3.2.2 Immunomodulation by Anaphylatoxins

In addition to their proinflammatory effects, both anaphylatoxins, C3a and C5a, have been reported to have immunomodulatory activity. However, in contrast to their similarity in mediating acute inflammation, their immunoregulatory effects are opposite, i.e., C5a was found to enhance and C3a to suppress immune responses (MORGAN et al. 1983, 1984). The anaphylatoxin receptors, C3aR and C5aR, belong to the seven transmembrane, G protein-coupled, rhodopsin family and are fully capable of signal transduction (GERARD and GERARD 1991; AMES et al. 1996). Both receptors are expressed on a wide range of cells, including human B cells, which express C3aR (FISCHER and HUGLI 1997) and human T cells which express C5aR (NATAF et al. 1999). However, the immunoregulatory effects of C3a and C5a are not apparently effected by signaling through their receptors on lymphocytes. Instead, they are attributed to effector molecules, mainly interleukins, secreted primarily by monocytes/macrophages upon stimulation by the anaphylatoxins. C5a enhances significantly secretion of (tumor necrosis factor) TNF, IL-1 (SCHINDLER et al. 1990), IL-6 (MORGAN et al. 1992), and IL-8 (EMBER et al. 1994) by lipopolysaccharide (LPS)- or IL-1β primed mononuclear cells. In addition, C5a suppresses production of IL-12 by IFN-γ/LPS-primed monocytes (WITTMANN et al. 1999). Similarly, C3a enhanced synthesis of TNF and IL-1β by LPS-stimulated adherent mononuclear cells (TAKABAYASHI et al. 1996) and of IL-6 by LPS- or IL-1β-stimulated mononuclear cells (FISCHER et al. 1999). The relevance of these in vitro observations to in vivo regulation of immune responses remains unknown.

4 Conclusions

Obviously our understanding of the contributions of complement to host defense has come a full circle. From a "factor" assisting antibodies in bacteriolysis to an important highly evolved element of innate immunity, to a system capable of pathogen recognition and thus, important for instructing adaptive immunity about targets and the magnitude of appropriate responses (Fearon and Locksley 1996).

Complement can recognize a large variety of pathogens either directly or indirectly through other host defense molecules. Encounter of pathogens triggers enzymatic reactions, which lead to the assembly of convertases cleaving C3 and C5 into biologically active fragments. A key event in complement activation is the attachment of C3b to the surface of complement activators through covalent bonds. C3b and its activation products iC3b, C3dg, and C3d express important functions including opsonin and immunoregulatory activity. Another important function is carried out by C5b, the initiator of the assembly of a large protein–protein complex, which can cause bacterial death. Finally, the two complement anaphylatoxins, C3a and C5a, contribute to defense against pathogens by mediating acute inflammatory reactions and perhaps also contributing to immunoregulation. Regulation of B cell responses by complement involves mainly antigen-linked C3d reacting with the CR2 receptor on B cells and FDCs and enhancing B cell activation and maturation along the B cell memory pathway.

References

Ahearn JM, Fischer MB, Croix D, Goerg S, Ma M, Xia J, Zhou X, Howard RG, Rothstein TL, Carroll MC (1996) Disruption of the *Cr2* locus results in a reduction in B-1a cells and in an impaired B cell response to T-dependent antigen. Immunity 4:251–262

Ames RS, Li Y, Sarau HM (1996) Molecular cloning and characterization of the human anaphylatoxin C3a receptor. J Biol Chem 271:20231–20234

Arya S, Chen F, Spycher S, Isenman D, Shulman MJ, Painter RH (1994) Mapping of amino acid residues in the Cμ3 domain of mouse IgM important in macromolecular assembly and complement-dependent cytolysis. J Immunol 152:1206–1212

Baumgarth N, Herman OC, Jager GC, Brown L, Herzenberg LA, Herzenberg LA (1999) Innate and acquired humoral immunities to influenza virus are mediated by distinct arms of the immune system. Proc Natl Acad Sci USA 96:2250–2255

Bianco C, Griffin FM, Silverstein SC (1975) Studies on the macrophage complement receptor: alteration of receptor function upon macrophage activation. J Exp Mad 141:1278–1290

Boettger EC, Metzger S, Bitter-Suermann D, Stevenson G, Kleindienst S, Burger R (1986) Impaired humoral immune response in complement C3-deficient guinea pigs: absence of a secondary immune response. Eur J Immunol 16:1231–1235
Bradbury LE, Kansas GS, Levy S, Evans RL, Tedder TF (1992) The CD19/CD21 signal transducing complex of human B lymphocytes includes the target of antiproliferative antibody-I and Leu-13 molecules. J Immunol 149:2841–2850
Bredt W, Wellek B, Brunner H, Loos M (1977) Studies on the interaction between *Mycoplasma pneumoniae* and the first component of complement. Infect Immun 15:7–12
Carroll MC (1998) The role of complement and complement receptors in induction and regulation of immunity. Annu Rev Immunol 16:545–568
Carroll MC, Prodeus AP (1998) Linkages of innate and adaptive immunity. Curr Opin Immunol 10:36–40
Carter RH, Fearon DT (1992) CD19: Lowering the threshold for antigen receptor stimulation of B lymphocytes. Science 256:105–107
Carter RH, Spycher MO, Ng YC, Hoffman R, Fearon DT (1988) Synergistic interaction between complement receptor type 2 and membrane IgM complex. J Immunol 141: 457–463
Clas F, Loos M (1981) Antibody-independent binding of the first component of complement and its subcomponent C1q to the S and R forms of *Salmonella minnesota*. Infect Immun 31:1138–1144
Cooper NR, Morrison DC (1978) Binding and activation of the first component of human complement by the lipid A region of lipopolysaccharides. J Immunol 120:1862–1868
Cooper NR, Jensen FC, Welsh RM, Oldstone MBA (1976) Lysis of RNA tumor viruses by human serum: direct antibody-independent triggering of the classical complement pathway. J Exp Med 144:970–984
Croix DA, Ahearn JM, Rosengard AM, Han S, Kelsoe G, Ma M, Carroll MC (1996) Antibody response to a T-dependent antigen requires expression of complement receptors. J Exp Med 183:1857–1864
Czop JK, Austen KF (1980) Functional discrimination by human monocytes between their C3b receptors and their recognition units for particulate activators of the alternative complement pathway. J Immunol 125:124–128
Daffern PJ, Pfeifer PH, Ember JA, Hugli TE (1995) C3a is a chemotaxin for human eosinophils but not for neutrophils. I. C3a stimulation of neutrophils is secondary to eosinophil activation. J Exp Med 181:2119–2127
Dempsey PW, Allison MED, Akkaraju S, Goodnow CC, Fearon DT (1996) C3d of complement as a molecular adjuvant: Bridging innate and acquired immunity. Science 271:348–350
Dierich MP, Ebenbichler CF, Marschang P (1993) HIV and human complement: mechanisms of interaction and biological implication. Immunol Today 14:435–439
Duncan AR, Winter G (1988) The binding site for C1q on IgG. Nature 332:738–740
Elsner J, Oppermann M, Czech W, Dobos G, Schopf E, Norgauer J, Kapp A (1994) C3a activates reactive oxygen radical species production and intracellular calcium transients in human eosinophils. Eur J Immunol 24:518–522
Ember JA, Jagels MA, Hugli TE (1998) Characterization of complement anaphylatoxins and their biological responses. In: Volanakis JE, Frank MM (eds) The human complement system in health and disease. Marcel Dekker, New York, pp 241–284
Ember JA, Sanderson SD, Hugli TE, Morgan EL (1994) Induction of interleukin-8 synthesis from monocytes by human C5a anaphylatoxin. Am J Pathol 144:393–403
Engel P, Zhou L-J, Ord DC, Sata S, Koller B, Tedder TF (1995) Abnormal B lymphocyte development, activation and differentiation in mice that lack or overexpress the CD19 signal transduction molecule. Immunity 3:39–50
Epstein J, Eichbaum Q, Sheriff S, Ezekowitz RAB (1996) The collectins in innate immunity. Curr Opin Immunol 8:29–35

Fang Y, Xu C, Fu Y-X, Holers VM, Molina H (1998) Expression of complement receptors 1 and 2 on follicular dendritic cells is necessary for the generation of a strong antigen-specific IgG response. J Immunol 160:5273–5279

Fearon DT (1998) The complement system and adaptive immunity. Sem Immunol 10: 355–361

Fearon DT, Austen KF (1997) Activation of the alternative complement pathway due to resistance of zymosan-bound amplification convertase to endogenous regulatory mechanisms. Proc Natl Acad Sci USA 74:1683–1687

Fearon DT, Locksley RM (1996) The instructive role of innate immunity in the acquired immune response. Science 272:50–54

Fischer WH, Hugli TE (1997) Regulation of B cell functions by C3a and $C3a^{desArg}$. J Immunol 159:4279–4286

Fischer WH, Jagels MA, Hugli TE (1999) Regulation of IL-6 synthesis in human peripheral blood mononuclear cells by C3a and $C3a_{desArg}$. J Immunol 162:453–459

Fischer MB, Goerg S, Shen LM, Prodeus AP, Goodnow CC, Kelsoe G, Carroll MC (1998) Dependence of germinal center B cells on expression of CD21/CD35 for survival. Science 280:582–585

Fischer MB, Ma M, Goerg S, Zhou X, Xia J, Finco O, Han S, Kelsoe G, Howard RG, Rothstein TL, Kremmer E, Rosen FS, Carroll MC (1996) Regulation of B cell response to T-dependent antigens by classical pathway of complement. J Immunol 157:549–556

Galdiero F, Tufano MA, Sommese L, Folgore A, Tedesco F (1984) Activation of complement system by porins extracted from *S. typhimurium*. Infect Immun 46:559–563

Gerard NP, Gerard C (1991) The chemotactic receptor for human C5a anaphylatoxin. Nature 349:614–617

Goetzl EJ, Austen KF (1974) Stimulation of human neutrophil leukocyte aerobic glucose metabolism by purified chemotactic factors. J Clin Invest 53:591–599

Goldstein IM, Kaplan HB, Radin A, Frosch M (1976) Independent effects of IgG and complement upon human polymorphonuclear leukocyte function. J Immunol 117: 1282–1287

Hamilton KK, Hattori R, Esmon CT, Sims PJ (1990) Complement proteins C5b-9 induce vesiculation of the endothelial plasma membrane and expose catalytic surface for assembly of the prothrombinase enzyme complex. J Biol Chem 265:3809–3814

Hebell T, Ahearn JM, Fearon DT (1991) Suppression of the immune response by a soluble receptor of B lymphocytes. Science 254:102–105

Heyman B, Wiesma EJ, Kinoshita T (1990) In vivo inhibition of antibody response by a monoclonal complement receptor specific antibody. J Exp Med 172:665–668

Ishiguro H, Kobayashi K, Suzuki M, Titani K, Tomonaga S, Kurosawa Y (1992) Isolation of a hagfish gene that encodes a complement component. EMBO J 11:829–837

Jackson CG, Ochs HD, Wedgewood RJ (1979) Immune response of a patient with deficiency of the fourth component of complement and systemic lupus erythematosus. N Engl J Med 300:1124–1129

Kantor AB, Herznberg LA (1993) Origin of murine B cell lineages. Annu Rev Immunol 11:501–538

Kishimoto TK, Jutila MA, Berg EL, Butcher EC (1989) Neutrophil Mac-1 and MEL-14 adhesion proteins inversely regulated by chemotactic factors. Science 245:1238–1241

Klaus GG, Humphrey JH (1977) The generation of memory cells. I. The role of C3 in the generation of B memory cells. Immunology. 33:31–40.

Klickstein LB, Wong WW, Smith JA, Weis JH, Wilson JG, Fearon DT (1987) Human C3b/C4b receptor (CR1). Demonstration of long homologous repeating domains that are composed of short consensus repeats characteristic of C3/C4 binding proteins. J Exp Med 165:1095–1112

Krop I, de Fougerolles AR, Hardy RR, Allison M, Schlissel MS, Fearon DT (1996) Self-renewal of B-1 lymphocytes is dependent on CD19. Eur J Immunol 26:238–242

Kurtz CB, O'Toole E, Christensen SM, Weis JH (1990) The murine complement receptor gene family. IV. Alternative splicing of *Cr2* gene transcripts predicts two distinct gene products that share homologous domains with both human CR2 and CR1. J Immunol 144:3581–3591
Li X, Carter RH (1998) Convergence of CD19 and B cell antigen receptor signals in the ERK2 activation cascade. J Immunol 161:5901–5908
Liszewski M, Farries T, Lublin D, Rooney I, Atkinson J (1996) Control of the complement system. Adv Immunol 61:201–283
Loos M, Wellk B, Thesen R, Opferkuch W (1978) Antibody-independent interaction of the first component of complement with gram-negative bacteria. Infect Immun 22:5–9
Marschang P, Kruger U, Ochsenbauer C, Gurtler L, Hittmair A, Bosch V, Patsch JR, Dierich MP (1997) Complement activation by HIV-1 infected cells: the role of transmembrane glycoprotein gp41. J Acquir Immune Defic Syndr Hum Retrovirol 14:102–109
Matsumoto AK, Kopicky-Burd J, Carter RH, Tuveson DA, Tedder TF, Fearon DT (1991) Intersection of the complement and immune systems: a signal transducing complex of the B lymphocyte-containing complement receptor type 2 and CD19. J Exp Med 173:55–64
Merino S, Nogueras MM, Aguilar A, Rubires X, Alberti S, Benedi VJ, Thomas JM (1998) Activation of the complement classical pathway (C1q binding) by mesophilic *Aeromonas hydrophila* outer membrane protein. Infect Immun 66:3825–3831
Molina H, Hollers VM, Li B, Fang Y-F, Mariathasan S, Goellner J, Strauss-Schoenberger J, Karr RW, Chaplin DD (1996) Markedly impaired humoral immune response in mice deficient in complement receptors 1 and 2. Proc Natl Acad Sci USA 93:3357–3361
Moore MD, Cooper NR, Tack BF, Nemerow GR (1987) Molecular cloning of the cDNA encoding the Epstein-Barr virus/C3d receptor (complement receptor type 2) of human B lymphocytes. Proc Natl Acad Sci USA 84:9194–9198
Morgan EL, Weigle WO, Hugli TE (1984) Anaphylatoxin-mediated regulation of human and murine immune responses. Fed Proc 43:2543–2547
Morgan EL, Thoman ML, Weigle WO, Hugli TE (1983) Anaphylatoxin-mediated regulation of the immune response. II. C5a-mediated enhancement of human humoral and T cell-mediated immune responses. J Immunol 130:1257–1261
Morgan EL, Sanderson S, Scholz W, Noonan DJ, Weigle WO, Hugli TE (1992) Identification and characterization of the effector region within human C5a responsible for stimulation of IL-6 synthesis. J Immunol 148:3937–3942
Nataf S, Davoust N, Ames RS, Barnum SR (1999) Human T cells express the C5a receptor and are chemoattracted to C5a. J Immunol 162:4018–4023
Nepomucenco RR, Henschen-Edman AH, Burgess WH, Tenner AJ (1997) cDNA cloning and primary structure analysis of C1qRp, the human C1q/MBL/SPA receptor that mediates enhanced phagocytosis in vitro. Immunity 6:119–129
Niculescu F, Rus H, van Biesen T, Shin ML (1997) Activation of Ras and mitogen-activated protein kinase pathway by terminal complement complexes is G protein dependent. J Immunol 158:4405–4412
Nonaka M (1998) Phylogeny of the complement system. In: Volanakis JE, Frank MM (eds) The human complement system in health and disease. Marcel Dekker, New York, pp 203–216
Nonaka M, Takahashi M (1992) Complete complementary DNA sequence of the third component of complement of lamprey. Implication for the evolution of thioester containing proteins. J Immunol 148:3290–3295
Ochs HD, Wedgewood RJ, Frank MM, Heller SR, Hosea SW (1983) The role of complement in the induction of antibody responses. Clin Exp Immunol 53:208–216
O'Neil KM, Ochs HD, Heller SR, Cork LC, Morris JM, Winkelstein JA (1988) Role of C3 in humoral immunity: defective antibody production in C3-deficient dogs. J Immunol 140:1939–1945

Papamichail M, Gutierez C, Embling P, Johnson P, Holborow EJ, Pepys MB (1975) Complement dependence of localization of aggregated IgG in germinal centers. Scand J Immunol 4:343–347

Pepys M (1972) Role of complement in induction of the allergic response. Nature New Biol 237:157–159

Pereira P, Forni L, Larsson E-L, Cooper M, Heusser C, Coutinho A (1986) Autonomous activation of B and T cells in antigen-free mice. Eur J Immunol 16:685–688

Plumb ME, Sodetz JM (1998) Proteins of the membrane attack complex. In: Volanakis JE, Frank MM (eds) The human complement system in health and disease. Marcel Dekker, Inc., New York, pp 49–82

Rickert RC, Rajewsky K, Roes J (1995) Impairment of T-cell-dependent B-cell responses and B-1 cell development in CD19-deficient mice. Nature 376:352–355

Roberts T, Snow EC (1999) Recruitment of the CD19/CD21 coreceptor to B cell antigen receptor is required for antigen-mediated expression of Bcl-2 by resting and cycling hen egg lysozyme transgenic B cells. J Immunol 162:4377–4380

Ross SC, Densen P (1984) Complement deficiency states and infection: epidemiology, pathogenesis and consequences of neisserial and other infections in an immune deficiency. Medicine 63:243–273

Saadi S, Platt JL (1995) Transient pertubation of endothelial integrity induced by antibodies and complement. J Exp med 181:21–31

Sastry K, Ezekowitz RA (1993) Collectins: pattern recognition molecules involved in first line host defense. Curr Opin Immunol 5:59–66

Sato S, Steeber DA, Tedder TF (1995) The CD19 signal transduction molecule is a response regulator of B-lymphocyte differentiation. Proc Natl Acad Sci USA 92:11558–11562

Schindler R, Lonnemann G, Hugli TE, Koch KM, Dinarello CA (1990) Transcription, not synthesis, of interleukin-1 and tumor necrosis factor by complement. Kidney Int 37:85–93

Smith LC, Chang L, Britten RJ, Davidson EH (1996) Sea urchin genes expressed in activated coelomocytes are identified by expressed sequence tags. Complement homologues and other putative immune response genes suggest immune system homology within the deuterostomes. J Immunol 156:593–602

Szalai AJ, Agrawal A, Greenhough TJ, Volanakis JE (1999) C-reactive protein: Structural biology and host defense function. Clin Chem Lab Med 37:265–270

Takabayashi T, Vannier E, Clark BD, Margolis NH, Dinarello CA, Burke JF, Gelfand JA (1996) A new biologic role for C3a and C3a desArg. Regulation of TNF-α and IL-1β synthesis. J Immunol 156:3455–3460

Tedder TF, Inaoki M, Sato S (1997) The CD19-CD21 complex regulates signal transduction thresholds governing humoral immunity and autoimmunity. Immunity 6:107–118

Tenner AJ, Robinson SL, Ezekowitz RAB (1995) Mannose binding protein (MBP) enhances mononuclear phagocyte function via a receptor that contains the 126000 Mr component of the C1q receptor. Immunity 3:485–493

Thyphronitis G, Kinoshita T, Inoue K, Schweinle JE, Tsokos GC, Metcalf ES, Finkelman FD, Balow JE (1991) Modulation of mouse complement receptors 1 and 2 suppresses antibody responses in vivo. J Immunol 147:224–230

Turner MW (1996) Mannose-binding lectin: the pluripotent molecule of the innate immune system. Immunol Today 17:532–540

Tuveson DA, Ahearn JM, Matsumoto AK, Fearon DT (1991) Molecular interactions of complement receptors on B lymphocytes: a CR1/CR2 complex distinct from the CR2/CD19 complex. J Exp Med 173:1083–1089

Volanakis JE (1982) Complement activation by C-reactive protein complexes. Ann NY Acad Sci 389:235–250

Volanakis JE (1990) Participation of C3 and its ligands in complement activation. Curr Top Microb Immunol 153:1–21

Volanakis JE (1998) Overview of the complement system. In: Volanakis JE, Frank MM (eds) The human complement system in health and disease. Marcel Dekker, New York, pp 9–32

Weis JJ, Toothaker LE, Smith JA, Weis JH, Fearon DT (1988) Structure of the human B lymphocyte receptor for C3d and the Epstein-Barr virus and relatedness to other members of the family of C3/C4 binding proteins. J Exp Med 167:1047–1066

Wittmann M, Zwirner J, Larsson V-A, Kirchhoff K, Begemann G, Kapp A, Götze O, Werfel T (1999) C5a suppresses the production of IL-12 by IFN-γ-primed and lipopolysacharide-challenged human monocytes. J Immunol 162:6763–6769

Several MHC-Linked Ig Superfamily Genes Have Features of Ancestral Antigen-Specific Receptor Genes

L. Du Pasquier

Abbreviations

A33	CTX-related human gene (Q99795), ref. for mapping (Johnstone 1999)
B-G	MHC-linked antigen on chicken hematopoietic cells
BEAT	*Drosophila* beaten path precursor
Buty	Butyrophilin
CAR	Coxsackie virus receptors (P78310)
CD4	Co-receptor CD1 with a first V domain (with an intron, but not a type 0 splice site as in CTX)
CD79a	Co-receptor of the B cell receptor formerly Igα or Mb1.Igsf C2 domain
CD83	Marker of a subset of dendritic cells
CD147	Basigin (Igsf member)
CEA	Carcinoembryonic antigen=CD66
CRAM1	New CTX-related molecule (Aurrand-Lions et al. 2000)

Basel Institute for Immunology, Grenzacherstrasse 487, 4005 Basel, Switzerland

CRTAM	Class-I MHC-restricted T cell-associated molecule (*Homo sapiens*). Human homologue of CTADS
CTADS	Chicken thymocyte activation and developmental protein; its first domain shows similarity with V domains of shark new antigen receptors. The second domain is close to C1
CTHumx	X-linked CTX human homologue (the closest for the external domains [Du Pasquier 2000a,b])
CTM.CTH	Mouse (m) and human (h) CTX-related molecules (Chrétien et al. 1998)
CTX	Cortical thymocyte marker of *Xenopus* (U43330), now also found in the digestive tract (Chrétien et al. 1998)
CTXR	CTX-related gene or sequence
EVA	Epithelial vascular antigen (AF030455). From the chicken data this region is also associated with the CD3 components (T. Goebel, personal communication)
FCGRT	Fc receptor IgG alpha chain transporter. Molecule of the MHC class I-type (NM_004107)
FREP	Fibrinogen-related proteins
FUT 1–7	Fucosyl transferases (Costache et al. 1997)
HCAR	Human coxsackie virus receptor
HTG	High throughput genomic sequences
Ig	Immunoglobulins
Igsf 1	Human sequence homologue of ILTs (AF034198)
Igsf 4	(NM014333) human Ig superfamily member with homology with PVR and NCAM
ILTs	Immunoglobulin-like transcripts (Nakajima et al. 1999)
Kappa	Ig kappa light chain
KIRs	Killer inhibitory receptors (reviewed in Lanier 1998)
LAG-3	Lymphocyte activation gene 3
Lambda	Ig lambda light chain
MAG	Myelin-associated glycoprotein (AAB58805)
MHC	Myosin heavy chain
MOG	Myelin oligodendrocyte glycoprotein (U18798)
MUC-18	Melanoma-associated protein (M28882)
NAR	New antigen receptor
NCAM	Neural cell adhesion molecule
NK-p30	Natural killer activating receptor=1C7 (AF031138)
NK-p44	Natural killer cell activating receptor (AJ225109)
P0	P0 myelin protein (V of the CTX type)
Poly Ig Receptor	Igsf members made of 5 V domains (X73079)

PreTα	Igsf C domain of the pre-T cell receptor
PreTα R	Genomic region homologous to a region situated upstream of the pre-T cell receptor alpha gene of chromosome 6 (AL035587)
PRR	Poliovirus receptor-related gene
PVR	Poliovirus receptors (M24406)
RAGE	Receptor for advanced glycolization end-product (Q15109)
SiRP	Signal regulatory protein
Tapas 9	(V2320 in ref) tapasin-related gene segment on chromosome 9
Tapas 11	(AC406 in ref) tapasin-related gene segment on chromosome 11
Tapas 12	(AC005840 in ref) tapasin-related gene segment on chromosome 12
Tapasch	Chicken tapasin
Tapashum	Human MHC-linked tapasin
Tapasin-R	Tapasin-related sequences
Tapaszf	Zebrafish tapasin
TCR	T cell receptor
TCRA	Human TCR alpha
TCRB	Human TCR beta
TCRDhum	Human TCR delta
TCRGhum	Human TCR gamma
TREM	Triggering receptor expressed on monocytes
(V0)	V domain with a type 0 splicing of the two half domain exons
V$2320	gene segment from chromosome 9 q34 related to V tapasin (Du Pasquier 2000a)
V$AC584ED	Tapasin chromosome 12p12 (Du Pasquier 2000a)
V$HS159	V domain, single exon, related to CTX and present on chromosome X q13
V$TAPAS11	V domain of the tapasin family found upstream of the C1 domain AC00406 (Du Pasquier 2000b)
V$TAPASCH	Chicken tapasin (AL023516)
V$TAPASHUM	Human tapasin (Y13582)
V$TAPASZF	Zebrafish tapasin (AAD41075)
V$ZFCTX1	A zebrafish CTX homologue
VH$HUM	Ig human variable heavy chain
Xist	X-inactive specific transcript
Xlp	X-linked lymphoproliferative disease

1 Introduction

The lymphocytes of the jawed vertebrates (gnathostomes) are characterized by their antigen-specific receptors. Both the Ig of B cells and the TCRs of T cells belong to the immunoglobulin superfamily. The antigen-binding domain of the variable part of these receptors is generated by the somatic rearrangement of gene segments. The introduction of this mechanism during evolution is likely to have occurred only once and must have had "rapid" consequences which shaped the immune system of vertebrates into a coherent coevolving unit (Marchalonis and Schluter 1990; Thompson 1995). This gives the impression of an abrupt "invention" of the whole adaptive immune system (review in Du Pasquier and Flajnik 1999). Yet the many elements of the immune system must have been acquired in a stepwise manner. For instance, antigen-presenting molecules of the class I- or class II-type have not appeared simultaneously under their recognizable form (Flajnik et al. 1991; Kaufman et al. 1984). Was the common ancestor of these types present before the introduction of somatic rearrangement? Or did it acquire its characteristics to match the need for selection created by somatic rearrangement? As far as the antigen receptor is concerned, somatic rearrangement mechanisms and the separation of the ABCDEF strands from the G strand in variable region genes must have also been introduced in pre-existing unsplit genes. Were these non-rearranging ancestral genes already committed to a lymphocyte-like lineage? Did they already encode molecules with an immune function? Where, in primitive genomes, were their genes located? Have they been duplicated many times?

The present chapter will describe MHC-linked Igsf genes and their relatives (some of them identified for the first time) on other linkage groups. Classifying this gene will perhaps help to answer some of the above questions and to facilitate the reconstruction of the pathway leading to the development of the vertebrate immune system.

2 Possible Structure of the Immediate Ancestor of an Antigen-Specific Receptor Gene: V-C1 Without Rearrangement

By comparison with Ig or TCR, a likely and close ancestor of the somatically rearranging receptor could have been a molecule with an extracellular segment comprising one variable (V) and one constant (C1) domain of the immunoglobulin superfamily. Non-rearranging V domains are ancient. They can be found in diploblastic invertebrates (in *Hydra* for instance, MILLER and STEELE, accession number HVU69448) or in Porifera (MÜLLER et al. 1999). They can also be easily recognized in various triploblastic invertebrates where they differ from the I set or C2 domains (in the case of *Drosophila* Amalgam, Lachesin and *Biomphalaria* FREPs for instance, review in DU PASQUIER and FLAJNIK 1999). C1 domains are less common and are probably latecomers (DU PASQUIER and CHRÉTIEN 1996). Although the I set has features of C1 domains, no C1 has been found in the entire genome of *Coenorhabditis elegans* (TEICHMANN and CHOTHIA 2000). So far they are found only in vertebrates that use the somatic rearrangement (i.e. the gnathostomes). Moreover, in the genome of vertebrates, their level of duplication, the number of their paralogues is so far not very high, as if they were a relatively recent acquisition. Yet they must have arisen from something and one of the purposes of this chapter is to trace the origin of C1 domains. The other purpose is to estimate how ancient they are and when they became associated with V domains.

3 Speculation About V-C1 Gene Segments and Their Localization in the MHC or in Paralogous Linkage Groups

If the V-C1 configuration is ancient there could be some visible signs of its early presence. So far they are absent in invertebrate phyla and in primitive vertebrate classes. Therefore, one way to proceed is to explore the genome of vertebrates (i.e. that of humans) for genes or gene segments that would give away the history of the V-C1 lineage. Those elements could be leftover pseudogenes, duplicated forms of recent or ancient origin. In an attempt to find out which of those were more ancient we shall oversimplify and assume: (1) that the intron came late, i.e. the original V domain genes had no intron, and (2) that when paralogues of a gene are found, this newly found gene is older than another gene for which no paralogues are found. In an initial scenario, finding four paralogues of the V-C1 configuration associated with already proven ancient syntenies could be an indication that this feature was ancient, coming from a genome corresponding to an early deuterostome or protostomian ancestor. Whereas finding only one paralogous genes could mean that this feature was more recent. In another nonideal scenario where no paralogues could be found, a consequence of gene loss, the whole approach might turn out to be quite problematic.

Interestingly, among the variety of genes that could code for molecules with the above-mentioned V-C1 configuration (for a review see Du Pasquier 2000a) some (tapasin and butyrophilin) can be found in the MHC or in MHC paralogous regions (Abi Rached et al. 1999). The other V-C1-C1 genes such as SIRPs or the V-C1-C2 similar to the PVRs are found on other linkage groups. Several other genes that could represent an intermediate between an ancient precursor and the immediate ancestor of Ig and TCR are also encoded in the MHC region. There are single V domains, V domains associated with the more primitive C2, or I set domains instead of a C1, and which are present in many adhesion molecules throughout the animal kingdom. The V genes of these molecules have different exon–intron organization compared to each other and to the rearranging receptor genes, which suggests, once more, that the ancestral V domain was an unsplit gene coding for all the strands of one domain. The association with the MHC may not be fortuitous. Perhaps during the evolution of the immune system the rearranging receptor genes developed initially from those ancestral V and V-C1 genes that were present in the MHC class III region at a time when it lacked class I and class II genes. Later, bona fide antigen-presenting molecules were necessary for the smooth functioning of a system where clonal expansion of possibly dangerous autoreactive clones had to be monitored. Table 1 and Fig. 1 show members of the Ig superfamily that have been found in the vicinity or within the MHC of different species.

Since our last estimation (Du Pasquier 2000b), some genes have been further charactesrized. The gene AC406, thought to be a chromosome 11 isolated C1 domain, caught our attention because of it relatedness to cold-blooded vertebrate class I C1 domains. The discovery of a V domain (Fig. 2) that could splice to it on the same HTG genomic fragment changes the relevance of this gene. It now resembles a light chain or the V-C1 core of tapasin (Fig. 3). Whether or not it is a paralogue reflecting an ancient genome duplication cannot be said. The two other tapasin related genes belong to an accepted MHC paralogous group. The homology searches for this AC406 gene confirm the apparent relationship between the C1 tapasin core and members of the PVR family in at least their anterior V-C portion (Du Pasquier 2000a). Astonishingly, an in silico experiment where artificially assembling the NK-p30 V domain to the V pre-T C domain linked both to the MHC, resulting in a "gene" with good homology to zebrafish tapasin, somewhat illustrated that genes in the MHC environment revolve around a single V-C1 type.

Table 1. Relationships among immunoglobulin superfamily members (Igsf) within the MHC

Igsf molecules	Species	Structure	Chromosomal location or linkage group	Other linked genes
B-G	*Gallus domesticus*	V.TM.Cy dimer	All in or linked to MHC 6 p	
Butyrophilin	*Homo sapiens*	V.TM.Cy B30.2		
CD83	*Homo sapiens*	V(0).TM.Cy		
CTX	*Xenopus laevis*	V(0).C2.TM.Cy		
MOG	*Homo sapiens*	V. TM.Cy		
NK-p30	*Homo sapiens*	V(0).TM.Cy		
RAGE	*Homo sapiens*	V(0). C2.C2.TM.Cy		
Tapasin	*Homo sapiens*	E.V.C1.TM.Cy		
Pre-Tα	*Homo sapiens*	C1.TM. Cy		
A33 (CTX family)	*Homo sapiens*	V(0). C2.C2.TM.Cy	1 q (by analogy with mapping (rat))	
Poly Ig Receptor	*Homo sapiens*	V.V.V.V.V.TM.Cy	1 q31-q41	MHC paralogue
Tapasin-R3 (V2320)	*Homo sapiens*	–.V.–.–.–.?	9 q34	MHC paralogue
Tapasin-R1	*Homo sapiens*	E. V.C1. TM.Cy	12 p12-p13	MHC paralogue + NK complex
LAG-3	*Homo sapiens*	V. C2.C2.C2.TM.Cy	12 p13	
CD147 (V0)	*Homo sapiens*	C2.V(0).TM.Cy	19 p13	MHC paralogue
Tapasin-R2 (AC000406)	*Homo sapiens*	–.V.C1.–	11 p14	
V1235 (V0) (CTX related)	*Homo sapiens*	–.V(0).–	11 q	
CRTAM	*Homo sapiens*	V.C1.TM.Cy	11 q22	
CTADS (CRTAM homol.)	*Gallus domesticus*	V.C1.TM.Cy	?	X linked lympho-proliferative disease
CTH (CTX family)	*Homo sapiens*	V(0).C2.TM.Cy	11 q23	CD3 γδε (in chicken)
MUC 18	*Homo sapiens*	V.C2.C2.TM.Cy	11 q23	
PRR	*Homo sapiens*	V.C1.C2.TM.Cy	11 q23-q24	
β2 microglobulin	*Homo sapiens*	C1	15 q21-q22.2	
PVR	*Homo sapiens*	V.C1.C2.TM.Cy	19 q13.2	KiRs, ILTs
SiRPS	*Homo sapiens*	V.C1.C1.TM.Cy	20	
CAR (CTX family)	*Homo sapiens*	V(0?).C2.C2.Cy	21 q11.2-q21	
CTH (CTX family)	*Homo sapiens*	V(0).C2.TM.Cy	21 q11.2-q21	
VHS 159 (CTX related)	*Homo sapiens*	–.V.–	X	Xist, Pre-Tα homologue
CTX-HUMX (CTX family)	*Homo sapiens*	V(0).C2.TM.Cy	X q22	
Igsf 1	*Homo sapiens*	ILTs	X q25	

Chromosome 6 contains, outside the MHC, several genes related to the V family. The natural killer cell activating receptor NK-p44, which can be 38% homologous to TCR V1 gamma regions, shows also some homology with a gene present on chromosome 6 (AL133404) that has some homology

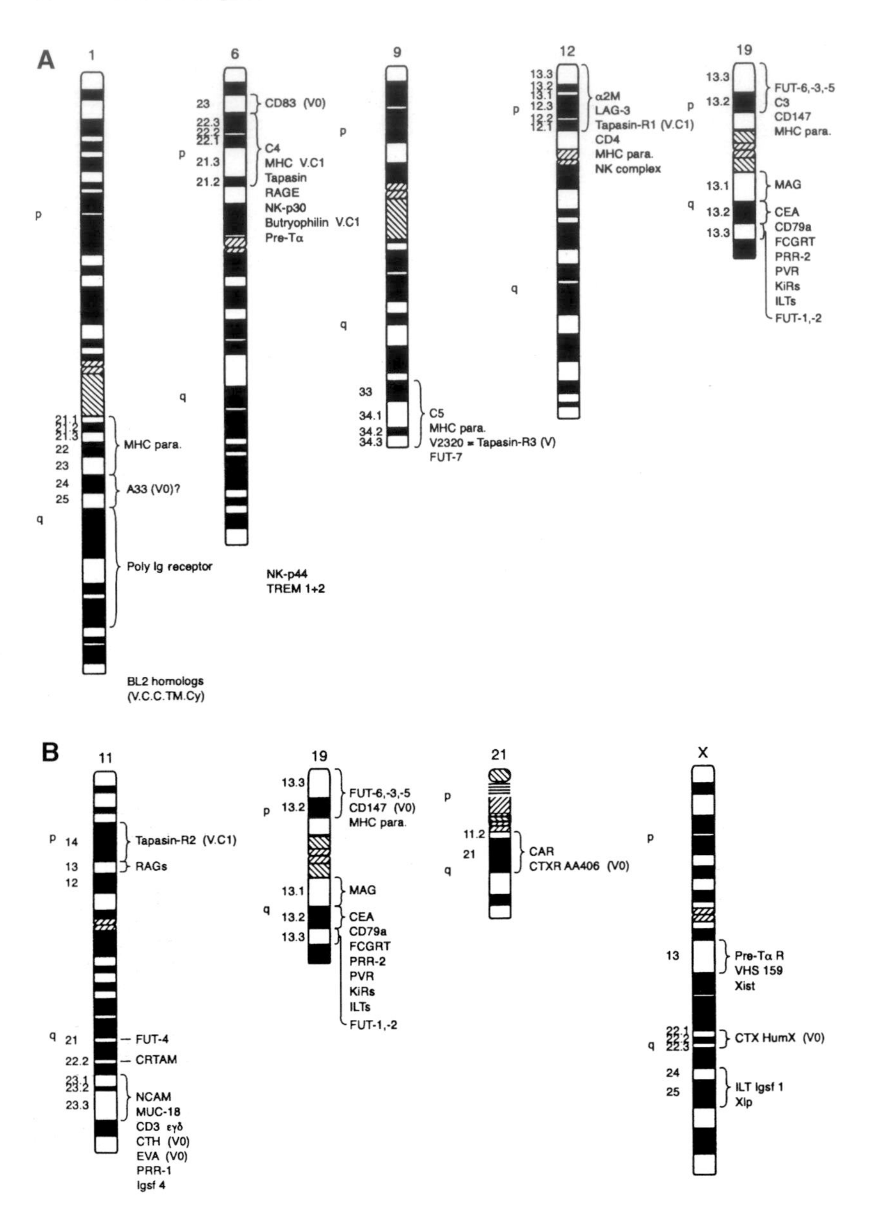

Fig. 1. A,B Chromosome localization of the genes described in Table 1

with the Poly Ig receptor of chromosome 1q (Figs. 1, 2). It turned out to be a fragment of the triggering receptor expressed on myeloid cells TREM2. It also shows 32% homology with TREM1 (Bouchon et al. 2000). It is not yet known whether or not these new genes form a cluster.

```
               : .          .                               : *       .
V$TAPASHUM VVLTVLTHTPAPRVRLGQDALLDLSFAYMPPTS--EAASSLAPGPPPFGLEWRRQHLGKG--HLLLAATPGLNGQ   71
   V$2320 PPSPVSPPPPANAEPSPERRRPQSVVGTRDPGPGSEMRTSWAQGPPPWPLAWQGLGGAKGA-RLSLRGADLLLAA   74
V$TAPASCH VALAVLTHTPTLRARVGSP--IHLHCAFAAP-------------PSSFVLEWRHQNRGAG--RVLLAYDSST-AR   57
V$TAPASZF VILSVSSKTPVVRCRLGEP--VLLDCGFWIDPSSP-------LHGSGFSIEWRYQFRGEG--RLVLAYDGKNDRF   64
V$AC584ED --FQVMTQTQSLSFLLGSS--ASLDCGFSMAP-----------GLDLISVEWRLQHKGRG--QLVYSWTAGQGQA   58
V$TAPAS11 -KVEMMAGG-TQITPLNDN--VTIFCNIFYSQP---------LNITSMGITWFWKSLTFDKEVKVFEFFGDHQEA   62
    ruler 1.......10........20........30........40........50........60........70.....

                                          *        . : . :         *  : *
V$TAPASHUM MPAAQEGAVAFAAWDDDEPWG-PWTGNGTFWLPRVQPFQEGTYLATIHLPYLQ--GQVTLELAVY-------  133
   V$2320 GPALGRLEGRHPGAENGG-------TVCFAALQPWAPWPGAGWHARLWLGWVQPLEQGTVVLGQL-------  132
V$TAPASCH APRAHPGAELLLGTRDGDG-----VTAVTLRLARPSPGDEGTYICSVFLPHGH--TQTVLQLHVFEPPKVTL  122
V$TAPASZF AETSESGAEMDITGLYQTG-------NASLILEESQVRHSGTYICTVYLPHLL--AQVAVDLEIV-------  120
V$AC584ED VRKGATLEPAQLGMARDA---------S-LTLPGLTIQDEGTYICQITTSLYR--AQQIIQLNIQ-------  111
V$TAPAS11 FRPGAIVSPWRLKSGDAS---------L--RLPGIQLEEAGEYRCEVVVTPLK--AQGTVQLEVV-------  114
    ruler ...80........90.......100.......110.......120.......130.......140.......
```

Fig. 2. Alignment of tapasin (V domain) related sequences. Clustal W alignment with gap penalty 15 and PAM series protein weight matrix

4 Does V-C1 or C1 Exist Outside of the MHC or MHC Paralogous Regions?

The β2 microglobulin is on chromosome 15q21, and Ig and TCR are on chromosome 2 (Igκ L chain is on 2p127); TCRγ is on 7p15; TCRβ is on 7q35,14 (Ig heavy chain on 14q32.3, TCR and on 14 p11.2,22); and Igλ L chain on 22p11.2. This lack of correlation with MHC must mean that the separation from the ancestral gene has taken place early in the history of the vertebrates, when the genome was not yet fixed in its multiduplicated form.

The SIRPs made of V-C1-C1 external domains (Adams et al. 1998) are on chromosome 20p13. SIRPs, present on chromosome 20 in human, bear some homology to the I set Ig domains of hemicentin, a *C. elegans* molecule. The homology is best in the V domain. A gene with a section homologue of the non-Ig part of hemicentin is present within the MHC class III region of the mouse (Albertella et al. 1996). This gene, G7c in mouse and NG37 in human, does not contain Ig domains; however, the homology of the anterior part is highly significant. During evolution there may have been interesting exon shuffling in an MHC-linked gene.

PVRs or PRRs with the first C domains closer to C1, rather than C2 or I set domains are on linkage group (chromosome 19 and 11) that contains many genes pertinent to the immune system. Tage4, a tumour-associated antigen (AF125555) is on rat 1q22, a region homologous to a section of the long arm of the human chromosome 19q. The specialists do not consider the q region of the chromosome 19 as a regular MHC-paralogous region. Yet the presence of the fucosyl transferase genes in MHC paralogous

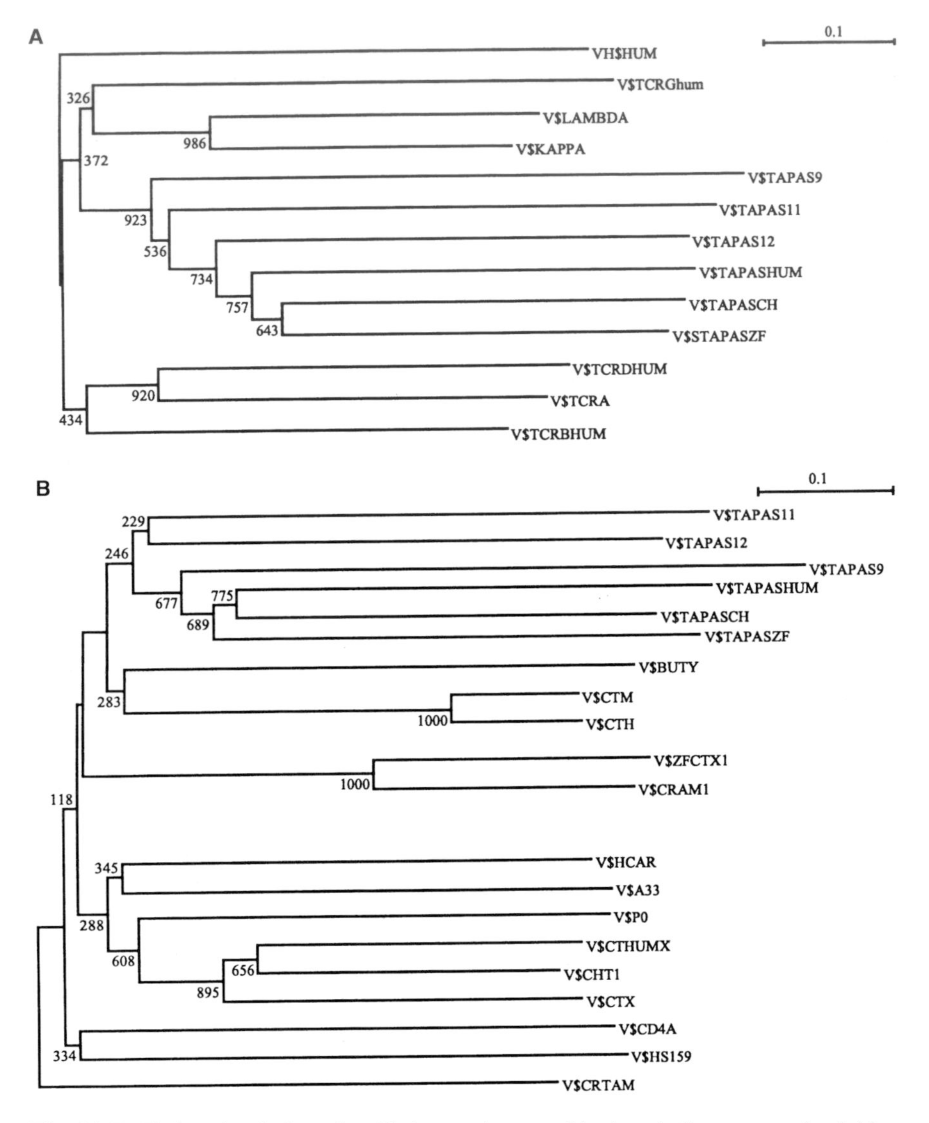

Fig. 3A,B. V domain phylogenies. Phylogenetic tree with clustal alignment and neighbour-joining algorithm for the tree. The dendrogram is meant to show grouping of sequences in families. Boostrap values given as an indication are poor. Yet the tapasin sequences (Bootstrap value of the whole group=923/1,000) stay grouped with each other in parts **A** and **B**. **A** Comparison of the tapasin-related V segment to Ig and TCR. **B** Comparison of the tapasin-related segments to the CTX family members, CD4 and CRTAM

regions 9q34 and 19p13 suggests a relationship between these two regions and 19q13 (Costache et al. 1997).

CRTAM, a gene found in the database, looks like a V-C1-TM-Cy (Du Pasquier 2000a,b) molecule and is encoded next to a human CTX homologue on chromosome 11q22 near another FUT gene in a region with homologies and synteny conservation with chromosome 19q13 [e.g. PVR-related sequences. Its homologous CTADS in chicken has now been reported to function in T cell activation and development (Ruble and Foster 2000)].

The first domain of CRTAM shows a strong relationship with the V domain of the shark Igsf Nar receptor, suggesting that, structurally speaking, the first domain of CRTAM may in fact be a V domain. CRTAM shows strong homology (30% over 185 amino acids) with BL2 (GenBank accession number AF132811) an Igsf member of the V-C2-C2 type and located in the MHC paralogous region on chromosome 1q22. BL2 is itself very homologous to a *Drosophila* Igsf member BEAT. The predicted structural domains of BEAT can be reasonably aligned with TCR Vα and Vβ sequences (Mushegian 1997).

5 NK Cells and Their Receptors

In a recent speculation, I had discarded the KIRs from being direct ancestors of the Ig and TCR on the basis that they did not have V domains. This does not mean that the NK cells, themselves, do not have other receptors that could be on the path leading to the TCR and Ig. Probably, NK cells still represent the best candidate for a cell type resembling a primordial lymphocyte. Perhaps they will turn out to be homologous to similar cell types found in some invertebrates (Pende et al. 1999). In our scheme, where MHC class I and class II comes late in evolution, KIRs that recognize the MHC would also be a relatively late acquisition. What is really primordial about the NK cell is its series of activating receptors. One of them, 1C7, recently found to correspond to the NK-p30 activating receptor, is a V domain encoded in the MHC. It has many interesting features. An alternate splicing can give its V domain some features of C2. This alternative splicing use a type 0 splice law as in CTX family members, a category of molecule also MHC linked in frogs and with some members, linked to MHC paralogues in human. NK-p44, another activating receptor with an Igsf, V domain is also on chromosome 6 but to our knowledge has not yet been mapped with precision (Cantoni et al. 1999). Interestingly, TREM1 and TREM2, two related receptors, although not expressed on NK cells but on monocytes and neutrophils and made of single Igsf V domains, also map to chromosome 6 (Bouchon et al. 2000, see above).

6 Conclusions

In mammalian genomes V-C1 gene segments are not as numerous as V-C2 or multiple C2 segments and they cannot be found in invertebrates. C1 domains found in Ig, TCR, MHC class I and II, tapasin, butyrophilin, β2 microglobulin, SIRPs, and CRTAM remain strictly associated with species that make use of the somatic rearrangement to generate their antigen specific receptor genes. For this reason, they look relatively modern. The immunoglobulin superfamily I set has features of both C1 and V domains. However, its evolution from this apparently primitive domain towards a real V seems to have taken place earlier than the evolution towards C1, since V domains, but not C1 can be found in invertebrates. The small number of real C1 paralogues also serves as argument for their recent development during the evolution of vertebrates. Some of these genes can be found in the MHC together with other gene segments that could also belong to the early lineage of the Igsf members that gave rise to the antigen-specific receptor.

Kasahara has proposed that the chromosomal duplications that formed the modern MHC paralogous group contributed to the emergence of several key accessory molecules of the adaptive immune system (1999). With the same reasoning it is also possible to propose that the MHC class III region contributed to the emergence of the antigen-specific receptors genes before the class I and class II were recruited in the same region. It is not yet possible to decide which came first, the antigen receptor, with somatic rearrangement, or the molecules than can effect the selection. Personally (and somewhat arbitrarily), I think that the more logical path follows the Darwinian principle according to which, variation precedes selection. This also applies to the creation of molecular species. Therefore, I presume that the introduction of somatic rearrangement which procured an enormous variation in the receptor binding sites preceded the recruitment of MHC class I and II that effect selection in the immune system. The early presence of V domains and the late appearance of the C1 domains typical of the modern class I and II molecules fit in with this sequence of events. One can still argue that MHC class I and II existed before "permitting" the somatic rearrangement, but so far, from the presence of "fossil" sequences, there is less evidence in the various genomes that have been analysed for this pathway. In both situations, the molecular similarity between the C1 domains as well the restriction of these molecular categories to gnathostomes argues for a rather rapid generation of these key elements within the vertebrate lineage.

More work should now been done to identify the origin of the peptide-binding region of the MHC class I and II molecules. This would complement the present one-sided view. Also more work will be necessary in invertebrate, early chordates and agnathans so that one can identify Ig superfamily linkage groups and determine whether the above predictions, namely finding genes homologous to 1C7, CTX or tapasin in a linkage group syntenic of MHC class III, can be fulfilled. More generally, in invertebrates it is not impossible that one can still find homologues of apparently more primordial receptors, for instance those belonging to the lineage that diverged early like CTX with the type 0 splice. More work should also have been done on the mode of binding of the Ig domains in those various molecules (tapasin for instance and CTX) to find out which function might have been selected in the evolution of the domains, since the Ig domain can clearly interact with various ligands in many ways (BORK et al. 1994). In this work, I considered the molecules in function of their linkage to MHC. There are other genes with V domain that do not rearrange somatically and which could be of interest (DU PASQUIER 2000a). These should be considered in a phylogenetic perspective. In light of it all, more questions remain unanswered. When was binding via the loops of the V domains selected? Was the binding to co-receptors instrumental in the selection of C1 domains? Indeed, when one looks at molecules with C1 domains, many bind to a co-receptor (CD3, CD4, CD8). Finally, the evolutionary aspects of the generation of cell lineage can be attempted in invertebrates via the isolation of the homologous transcription factors of the families involved in the generation of hematopoietic tissues in vertebrates (ANDERSON and ROTHENBERG 2000; HANSEN and MCBLANE 2000). This, together with the study of surface receptors, could one day lead to the understanding of relationships between NK cells and similar cell types in invertebrates (BOILEDIEU and VALEMBOIS 1977).

Acknowledgements. I thank Marco Colonna for critical reading of the manuscript and Allison Dwileski for help in the preparation of the figures and the manuscript. The Basel Institute for Immunology was founded and supported by F. Hoffman-La Roche Ltd., Basel, Switzerland.

References

Abi Rached L, McDermott MF, Pontarotti P (1999) The MHC big bang. Immunol Rev 167:33–44

Adams S, van der Laan LJ, Vernon-Wilson E, Renardel de Lavalette C, Dopp EA, Dijkstra CD, Simmons DL, van den Berg TK (1998) Signal-regulatory protein is selectively expressed by myeloid and neuronal cells. J Immunol 161:1853–1859

Albertella MR, Jones H, Thomson W, Olavesen MG, Campbell RD (1996) Localization of eight additional genes in the human major histocompatibility complex, including the gene encoding the casein kinase II beta subunit (CSNK2B). Genomics 36:240–251

Anderson MK, Rothenberg EV (2000) Transcription factor expression in lymphocyte development: clues to the evolutionary origins of lymphoid cell lineages? Curr Top Microbiol Immunol 248:137–155

Aurrand-Lions MA, Duncan L, Du Pasquier L, Imhof BA (2000) Cloning of JAM-2 and JAM-3: an emerging junctional adhesion molecular family. Curr Top Microbiol Immunol 251:91–98

Boiledieu D, Valembois P (1977) Natural cytotoxic activity of sipunculid leukocytes on allogenic and xenogenic erythrocytes. Dev Comp Immunol 1:207–216

Bork P, Holm L, Sander C (1994) The immunoglobulin fold. Structural classification, sequence patterns and common core. J Mol Biol 242:309–320

Bouchon A, Dietrich J, Colonna M (2000) Cutting edge: inflammatory responses can be triggered by TREM-1, a novel receptor expressed on neutrophils and monocytes. J Immunol 164:4991–4995

Cantoni C, Bottino C, Vitale M, Pessino A, Augugliaro R, Malaspina A, Parolini S, Moretta L, Moretta A, Biassoni R (1999) NKp44, a triggering receptor involved in tumor cell lysis by activated human natural killer cells, is a novel member of the immunoglobulin superfamily. J Exp Med 189:787–796

Chrétien I, Marcuz A, Courtet M, Katevuo K, Vainio O, Heath JK, White SJ, Du Pasquier L (1998) CTX, a *Xenopus* thymocyte receptor, defines a molecular family conserved throughout vertebrates. Eur J Immunol 28:4094–4104

Costache M, Cailleau A, Fernandez-Mateos P, Oriol R, Mollicone R (1997) Advances in molecular genetics of alpha-2- and alpha-3/4-fucosyltransferases. Transfus Clin Biol 4:367–382

Du Pasquier L (2000a) The phylogenetic origin of antigen-specific receptors. Curr Top Microbiol Immunol 248:160–185

Du Pasquier L (2000b) Relationships among the genes encoding MHC molecules and the specific antigen receptors. In: Kasahara M (ed) Major histocompatibility complex, evolution structure, and function. Springer, Tokyo. 53–65

Du Pasquier L, Chrétien I (1996) CTX, a new lymphocyte receptor in *Xenopus* and the early evolution of Ig domains. Res Immunol 147:218–226

Du Pasquier L, Flajnik MF (1999) Origin and evolution of the vertebrate immune system. In: Paul WE (ed) Fundamental immunology. Lippincott-Raven, Philadelphia

Flajnik MF, Canel C, Kramer J, Kasahara M (1991) Which came first, MHC class I or class II? Immunogenetics 33:295–300

Hansen JD, McBlane JF (2000) Recombination-activating genes, transposition, and the lymphoid-specific combinatorial immune system: a common evolutionary connection. Curr Top Microbiol Immunol 248:111–135

Johnstone CN (1999) Regulation of tissue specific expression of the A33 antigen. PhD, University of Melbourne

Kasahara M (1999) The chromosomal duplication model of the major histocompatibility complex. Immunol Rev 167:17–32

Kaufman JF, Auffray C, Korman AJ, Shackelford DA, Strominger J (1984) The class II molecules of the human and murine major histocompatibility complex. Cell 36:1–13

Lanier LL (1998) NK cell receptors. Annu Rev Immunol 16:359–393

Marchalonis JJ, Schluter SF (1990) On the relevance of invertebrate recognition and defence mechanisms to the emergence of the immune response of vertebrates. Scand J Immunol 32:13–20

Müller WE, Blumbach B, Müller IM (1999) Evolution of the innate and adaptive immune systems: relationships between potential immune molecules in the lowest metazoan phylum (Porifera) and those in vertebrates. Transplantation 68:1215–1227

Mushegian AR (1997) The *Drosophila* Beat protein is related to adhesion proteins that contain immunoglobulin domains [letter]. Curr Biol 7:R336–338

Nakajima H, Samaridis J, Angman L, Colonna M (1999) Human myeloid cells express an activating ILT receptor (ILT1) that associates with Fc receptor gamma-chain. J Immunol 162:5–8

Pende D, Parolini S, Pessino A, Sivori S, Augugliaro R, Morelli L, Marcenaro E, Accame L, Malaspina A, Biassoni R, Bottino C, Moretta L, Moretta A (1999) Identification and molecular characterization of NKp30, a novel triggering receptor involved in natural cytotoxicity mediated by human natural killer cells. J Exp Med 190:1505–1516

Ruble DM, Foster DN (2000) Molecular characterization of a new member of the immunoglobulin superfamily that potentially functions in T-cell activation and development. Immunogenetics 51:347–357

Teichmann SA, Chothia C (2000) Immunoglobulin superfamily proteins in *Caenorhabditis elegans*. J Mol Biol 296:1367–1383

Thompson CB (1995) New insights into V(D)J recombination and its role in the evolution of the immune system. Immunity 3:531–539

Coat Protein Transgenic Papaya: "Acquired" Immunity for Controlling Papaya Ringspot Virus

D. Gonsalves

1 Introduction

Pathogen-derived resistance (PDR) (Sanford and Johnston 1985) is an example of "acquired" immunity in plants in the sense that transgenic plants which express a viral gene are often resistant to that particular virus. This brief review examines the practical application of PDR to control papaya ringspot virus (PRSV) in Hawaii, arguably the most important disease of papaya worldwide (Gonsalves 1998; Gonsalves et al. 1998).

Department of Plant Pathology, Cornell University, New York State Agricultural Experiment Station, Geneva, NY 14456-0462, USA

2 The Papaya and Papaya Ringspot Virus

Papaya (*Carica papaya*) is a fruit crop that is widely grown in the lowlands of tropical and subtropical regions (Manshardt 1992). Its delicious flavor, high vitamin C and A content, and the relatively short time it takes for a tree to bear mature fruit makes it a common backyard fruit, as well as a commercially grown fruit crop. The Hawaiian solo papaya, because of it sweetness and good shipping qualities, is a favorite export papaya crop. Brazil and Jamaica export solo papaya to Europe while Hawaii supplies much of the solo papaya that is consumed in the USA, Canada, and Japan.

However, PRSV has limited the economic production of papaya in many areas of the world (Purcifull et al. 1984; Gonsalves 1994, 1998). The lack of resistant cultivars combined together with the rapid spread of the virus by aphids make PRSV difficult to control. Only marginal success in controlling PRSV have been achieved by planting papaya in isolation, diligent rouging of infected trees, cross protection, and breeding efforts to obtain PRSV-tolerant papaya lines (Yeh and Gonsalves 1994; Gonsalves 1998). None of these methods has given satisfactory results on a large scale.

3 Rationale for Developing Transgenic Papaya to Control Papaya Ringspot Virus in Hawaii

PRSV was first reported in Hawaii in the 1940s on Oahu Island, where the papaya industry was located at that time (e.g., Gonsalves 1998). In little over a decade, PRSV severely affected papaya orchards on Oahu Island. The industry relocated in the Puna district, the region of Hawaii's active volcanoes, on Hawaii Island where large tracts of relatively inexpensive lease land were available, rainfall and sunshine were consistent, and PRSV was not present. The industry flourished; and Puna was producing 95% of Hawaii's papaya by the 1980s. However, PRSV became a potential threat when it was discovered only about 19 miles away in the city of Hilo in the 1970s. Geographic isolation, constant virus surveillance, and prohibiting the introduction of papaya seedlings into Puna had kept the virus from Puna. Nevertheless, since it was very probable that PRSV would eventually enter the Puna district, research was started in 1978 to develop control strategies for PRSV in Hawaii. Cross protection was investigated as a control measure. A mild nitrous acid mutant, designated PRSV HA

5–1 (Yeh and Gonsalves 1984), was isolated and used to some extent on Oahu Island (Mau et al. 1989) and in Taiwan (Yeh and Gonsalves 1994). However, the mild mutant was not widely used because its effectiveness was limited to closely related isolates (Tennant et al. 1994; Yeh and Gonsalves 1994) and because it induced significant fruit symptoms on certain papaya cultivars (unpublished observations). It still is used, however, to cross protect papaya in Guam (G. Wall, personal communication).

4 The Concept of Pathogen-Derived Resistance and Development of Transgenic Papaya for Resistance to Papaya Ringspot Virus

In the early to mid-1980s, the concept of PDR was put forth and experimental evidence showed its promise for controlling viruses. Sanford and Johnston (1985) proposed the concept in 1985. It stated that expressing a transgene of a pathogen in a plant would result in the transgenic plant being resistant to that particular pathogen. Independently and about the same time, Abel (Powell-Abel et al. 1986) from Roger Beachy's group reported that transgenic tobacco expressing the coat protein gene of tobacco mosaic virus showed significant delay in disease symptoms caused by tobacco mosaic virus. These major advancements prompted our laboratory to explore, starting in 1987, the possibility of using PDR to control PRSV by transforming papaya with the coat protein gene of PRSV HA 5–1, a mild nitrous-acid mutant derived from the severe PRSV HA (Gonsalves and Ishii 1980; Yeh and Gonsalves 1984). We wanted to develop a plant with "acquired" immunity against the attacking virus.

The various steps that our group, which consisted of Richard Manshardt and Maureen Fitch of the University of Hawaii, Jerry Slightom of the Upjohn Company and myself, went through to develop the transgenic papaya was recently reviewed (Gonsalves 1998; Gonsalves et al. 1998). The following is a brief recap of events. The coat protein gene of PRSV HA 5–1 isolate was engineered for transformation into papaya. PRSV HA 5–1 was used because it was the first PRSV isolate that we had cloned in our laboratory and because it originated from a Hawaiian strain of PRSV. A major breakthrough was the regeneration of papaya from somatic embryos of calli that originated from zygotic embryos of immature papaya fruit (Fitch and Manshardt 1990). The gene gun was used to transform papaya because we had ready access to the gene gun at Geneva and had the help of John Sanford, a co-inventor of the technology (e.g., Sanford et al.

1992). The coat protein gene was under the control of a single 35S promoter and was engineered to express a translatable coat protein gene construct.

5 The Year 1992: Merging of a Research Program and a Disease Crisis

Transformation experiments resulted in the successful regeneration of nine transgenic papaya lines that were screened for resistance to PRSV HA (Fitch et al. 1992). Six transformants were of the Sunset cultivar and three of the Kapoho cultivar. Sunset, a commercial red-fleshed Hawaiian solo papaya, is widely grown in Brazil, but not so much in Hawaii. Kapoho, a yellow-flesh Hawaiian solo papaya, is the dominant cultivar grown in Hawaii. By 1991, we had tested and shown that one transgenic Sunset line, designated 55-1, was resistant to greenhouse inoculations of PRSV HA (Fitch et al. 1992). A field trial was started in April 1992 on Oahu island, using R0 cloned plants of the resistant line (Lius et al. 1997). By December 1992, results clearly showed that line 55-1 was resistant to PRSV in Hawaii under heavy virus pressure. Coincidentally, PRSV-infected papaya was discovered in the Puna district in May 1992. Thus, 1992 marked the beginning of a potential disaster to the Hawaiian papaya industry and the time when we obtained initial evidence that the transgenic papaya could control PRSV in Hawaii.

The PRSV resistance of transgenic papaya line 55-1 and the devastating effect of PRSV in Puna became clearer by 1995. Data from the 1992 field trial convinced us that the transgenic papaya was indeed highly resistant to PRSV under field conditions. In fact, by the end of 1992 nearly all of the nontransgenic papaya were infected while none of the R0 clones of line 55-1 showed symptoms (Lius et al. 1997). In contrast, PRSV moved through the Puna district much faster than expected and by the latter part of 1994, PRSV was widespread in Puna and the Hawaii Department of Agriculture had abandoned efforts to contain the virus by cutting down infected plants. Hawaii's papaya industry was in a crisis (Gonsalves 1998).

Transgenic plants and their products are regulated by APHIS (Animal Plant Health Inspection Service), EPA (Environmental Protection Agency), and by FDA (Food and Drug Administration), the latter being through a consultative process. APHIS assesses the environmental safety of the transgenic product. EPA regarded the coat protein as a pesticide and thus we needed to obtain an exemption tolerance levels of coat protein in transgenic plants. And, FDA assesses the food safety of the transgenic product. With the papaya crisis getting worse, we moved to satisfy conditions for having the transgenic papaya deregulated and commercialized. As a first step, a large field trial was starting in Puna in October 1995 using the cultivars UH SunUp (SunUp) and UH Rainbow (Rainbow) which had been developed from line 55-1 (MANSHARDT 1998). SunUp is line 55-1 that is homozygous for the coat protein gene. As mentioned above, line 55-1 was derived from Sunset which is a red-fleshed Hawaiian solo variety. Rainbow is an F1 hybrid of a cross of SunUp and Kapoho. Rainbow was preferred by Hawaiian papaya growers because of its yellow flesh and because it has properties of Kapoho, which was the dominant papaya cultivar grown in Hawaii.

The 1995 field trial clearly showed that SunUp and Rainbow were resistant to PRSV under severe disease pressure and evaluation of the fruit by horticulturist, growers, and packers also showed that these cultivars, especially Rainbow, were indeed a suitable substitute for the Kapoho cultivar (FERREIRA et al. 2001). In fact, test results showed that only three Rainbow plants out of over a 1,000 transgenic Rainbow became infected during the three-year trial. Rainbow plants produced estimated annualized yields of 100,000 pounds of marketable fruit per acre, while the non-transgenic papaya, all of which became infected, averaged about 100 times less marketable fruit.

By May 1998, the papaya was deregulated by APHIS and EPA, and we had completed the consultative process with FDA (GONSALVES 1998). Licenses to the intellectual property rights areas were obtained and seeds were distributed at no cost to farmers, and even homeowners, starting on 1 May 1998. Hawaii papaya growers had commercial PRSV-resistant transgenic cultivars to use in their fight against the disease that was devastating papaya in the Puna region.

6 The Year 2000: Transgenic Papaya in Puna

The transgenic Rainbow papaya is now widely planted in Puna, with many formerly abandoned fields being replanted with Rainbow. By all accounts, Rainbow has remained resistant to PRSV and has produced high yields of marketable fruit. In fact, papaya production increased in 1999, the first increase since 1993. The increase is largely due to the planting of Rainbow in the Puna area. From the technological standpoint, the transgenic papaya has been a success in controlling PRSV in Hawaii.

The years 1991 to even 1998 did not see much controversy over genetically engineered plants (called GMOs) in the USA. Furthermore, our efforts in developing the transgenic papaya were transparent, the Hawaiian papaya industry was in trouble and needed help, people (consumers) were aware of our efforts to solve the problem, and indeed people were happy when we did come up with a viable solution. After all, papaya is Hawaii's second most important commercial fruit crop. However, the GMO controversy that started in Europe is beginning to be felt in the USA and other parts of the world. Will this controversy affect the transgenic papaya in Hawaii and elsewhere?

Since Hawaii exports about 35% of its papaya to Japan, obtaining clearance to introduce transgenic papaya into Japan, and it subsequent acceptance by consumers is crucial to Hawaii's papaya industry. Hawaii also exports smaller amounts of papaya to Canada. Efforts are being made to gain clearance for the importation and sale of the transgenic papaya in Japan and Canada. It is hoped that these clearances will be obtained in the near future. Will the transgenic papaya be accepted in Japan and Canada, especially in the wake of the controversy over GMOs that is brewing in Europe? Time will tell. It is hoped that the implementation of this technology will not be unduly affected by the largely unfounded controversy on the safety of GMO products.

7 Nature of the "Acquired" Immunity in Transgenic Papaya

In 1987, when we first started the transgenic papaya work, coat protein-mediated protection was the prevailing hypothesis on the mechanism of PDR for plant viruses. However, largely through the pioneering work of the Bill Dougherty's group (LINDBO and DOUGHERTY 1992a,b; LINDBO et al. 1993), the hypothesis of RNA-mediated resistance emerged around 1993. Further work showed that post-transcriptional gene silencing (PTGS) was the underlying mechanism of RNA-mediated resistance.

Numerous works, especially from David Baulcombe's laboratory, have shown that PTGS is the underlying mechanism RNA-mediated resistance, and it appears that RNA-mediated resistance is much more prevalent then protein-mediated resistance. (BAULCOMBE 1996, 1999). This form of resistance is also called homology-dependent resistance since resistance is limited to viruses with high sequence homology to the viral transgene. The prevailing consensus is that PTGS is a plant mechanism for regulating levels of cellular RNAs and for protecting plants against viral pathogens.

A recent remarkable revelation is that plant viruses themselves can mount counter measures that overcome or lessen the effects of PTGS in plants (ANANDALAKSHMI et al. 1998; BECLIN et al. 1998; BRIGNETI et al. 1998; KASSCHAU and CARRINGTON 1998). Several laboratories have shown that transgenic plants that express specific viral proteins from potyviruses (ANANDALAKSHMI et al. 1998; KASSCHAU and CARRINGTON 1998) and cucumoviruses (BECLIN et al. 1998; BRIGNETI et al. 1998) do not display PTGS of various transgenes. In effect, the general defensive mechanisms (i.e., PTGS) of plants against plant viruses can be countered by certain plant viruses that apparently produce proteins that shutdown or diminish PTGS of plants. Recent works have shown that small RNA molecules, about 25bp in length, play an important role in the process of PTGS (HAMILTON and BAULCOMBE 1999).

RNA-mediated protection and PTGS were not known when we first identified the resistant transgenic papaya in 1991. However, subsequent work has shown that the resistance of the transgenic papaya is RNA-mediated and is due to PTGS. Also, coat protein gene dosage and plant developmental stage affect the resistance of the transgenic papaya; these are characteristics of RNA-mediated resistance and PTGS (BAULCOMBE 1996). Rainbow (which is hemizygous for the coat protein gene) is only resistant to isolates with coat protein genes that have very high homology (greater than 96%) to the transgene, while homozygous SunUp is resistant to isolates with lower coat protein homology (TENNANT et al. 1994; TENNANT 1996). The lower limit of homology to which SunUp is resistant is about 90%. For example, SunUp is resistant to the Brazil isolate of PRSV (93% homology) while it is susceptible to isolates from Thailand and Taiwan, which show 89%–90% homology. In relation to the plant developmental stage, Rainbow can be susceptible even to very closely related PRSV isolates when inoculated at a young stage, whereas they are resistant when inoculated at an older stage (TENNANT 1996).

8 Implementation of Transgenic Papaya Throughout the World

Since PRSV affects papaya worldwide, the PDR strategy employed to control PRSV in Hawaii could be used in other countries (GONSALVES 1998). However, two factors suggested that Rainbow and SunUp would not be suitable for controlling PRSV worldwide. First, the narrow resistance of Rainbow limits its usefulness to Hawaii, while SunUp would be effective in a number of countries, but not in Thailand and Taiwan, for example. Second, since many countries do not grow the Hawaiian solo papaya, resistance would need to be incorporated into cultivars that are acceptable to the target countries.

We are working to transfer the technology to control PRSV to other countries by developing transgenic papaya with coat protein genes of PRSV isolates from the target countries. We have transformed papaya with coat protein genes of PRSV isolates from Thailand, Brazil, Jamaica, and Venezuela. In general, these transgenic papaya are resistant to the isolates that supply the transgene but are largely susceptible to other isolates. Some lines of transgenic papaya do show a wider range of resistance but this is not generally the rule. From the practical standpoint, it would be best to transform plants with the coat protein gene of PRSV isolates from the target country.

We have developed a recent strategy that might result in transgenic papaya that have resistance to a range of PRSV isolates throughout the world. We recently showed that short segments of a viral coat protein gene will impart resistance if it is part of a chimeric gene that is long enough to induce PTGS (PANG et al. 1997). We have used this information to develop multiple-resistant transgenic plants. For example, transgenic *Nicotiana benthamiana* expressing a transgene composed of a full-length coat protein gene of turnip mosaic virus linked to a fragment of the nucleocapsid protein gene of tomato spotted wilt tospovirus show resistance to both viruses (JAN et al. 2000). In fact, resistance to three tospoviruses was observed in transgenic *N. benthamiana* with a chimeric transgene consisting of 200-bp segments (the nucleocapsid gene is ca. 800bp) of these three tospoviruses (JAN 1998).

Thus, transgenic papaya with resistance to PRSV isolates from Thailand, Taiwan, and Hawaii could be produced by transforming papaya with a single chimeric gene consisting of linked fragments of the coat protein genes of these PRSV isolates. A major advantage of this approach is that the resistance to the three PRSV strains would segregate as a single gene, which makes it convenient to transfer the multiple strain resistance to other cultivars through breeding. Of course, it might not be practical or even possible to engineer a papaya that has resistance to all PRSV isolates. However, developing resistance to isolates in a country and to isolates that might likely be introduced into the country would be an effective strategy. The sequence diversity of the isolates in the country as well as the surrounding countries would need to be known. The coat protein genes of many PRSV isolates in the world have been sequenced (e.g., BATESON et al. 1994; WANG and YEH 1997).

9 Final Comments: an Interface Between "Innate" and "Acquired" Immunity

Pathogen-derived resistance, a form of "acquired" immunity, has been successfully used to control PRSV in papaya (Gonsalves 1998). Interestingly, it appears that the mechanism of PDR for plant viruses is PTGS, a system used by plants for limiting or regulating the accumulation of specific RNAs. Thus, in one sense these plants have an "innate" immunity mechanism for resistance to plant viruses. This "innate" immunity then turns into "acquired" immunity when a viral transgene is introduced into a plant and the plant mobilizes PTGS which subsequently targets and degrades the homologous RNA of the attacking virus.

References

Anandalakshmi R, Pruss GJ, Ge X, Marathe R, Mallory AC, Smith TH, Vance VB (1998) A viral suppressor of gene silencing in plants. Proc Natl Acad Sci USA 95:13079–13084

Bateson M, Henderson J, Chaleeprom W, Gibbs A, Dale J (1994) Papaya ringspot potyvirus: isolate variability and origin of PRSV type P (Australia). J Gen Virol 75:3547–3553

Baulcombe DC (1996) Mechanisms of pathogen-derived resistance to viruses in transgenic plants. Plant Cell 8:1833–1844

Baulcombe D (1999) Viruses and gene silencing in plants. Arch Virol [Suppl] 15:189–201

Beclin C, Berthome R, Palaqui JC, Tepfer M, Vaucheret H (1998) Infection of tobacco or *Arabidopsis* plants by CMW counteracts systemic post-transcriptional silencing on non-viral (Trans)genes. Virology 252:313–317

Brigneti G, Voinnet O, Li WX, Ji LH, Ding SW, Baulcombe DC (1998) Viral pathogenicity determinants are suppressors of transgene silencing in *Nicotiana benthamiana*. Embo J 17:6739–6746

Ferreira SA, Pitz KY, Manshardt R, Zee F, Fitch M, Gonsalves D (2001) Coat protein transgenic papaya provides practical control of papaya ringspot virus in Hawaii. Plant Disease (in press)

Fitch M, Manshardt R (1990) Somatic embryogenesis and plant regeneration from immature zygotic embryos of papaya (*Carica papaya* L.). Plant Cell Rep 9:320–324

Fitch MMM, Manshardt RM, Gonsalves D, Slightom JL, Sanford JC (1992) Virus resistant papaya derived from tissues bombarded with the coat protein gene of papaya ringspot virus. Bio Technol 10:1466–1472

Gonsalves D (1994) Papaya ringspot virus. In: Ploetz R, Zentmyer G, Nishijima W, Rohrbach K, Ohr H (eds) Compendium of Tropical Fruit Diseases. APS Press, St. Paul, pp 67–68

Gonsalves D, Ishii I (1980) Purification and serology of papaya ringspot virus. Phytopathology 70:1028–1032

Gonsalves D (1998) Control of papaya ringspot virus in papaya: a case study. In: Webster RK, Shaner G, Van Alfen NK (eds) Annual Review of Phytopathology. Annual Reviews, Palo Alto, pp 415–437

Gonsalves D, Ferreira S, Manshardt R, Fitch M, Slightom J (1998) Transgenic virus resistant papaya: new hope for control of papaya ringspot virus in Hawaii. APSnet feature story for September 1998 on World Wide Web. Address is: http://www.scisoc.org/feature/papaya/Top.html

Hamilton AJ, Baulcombe DC (1999) A species of small antisense RNA in posttranscriptional gene silencing in plants. Science (Washington DC) 286:950–952

Jan F-J (1998) Roles of nontarget DNA and viral gene length. PhD Dissertation in Department of Plant Pathology, Ithaca, Cornell University, 286 pp

Jan FJ, Fagoaga C, Pang SZ, Gonsalves D (2000) A single chimeric transgene derived from two distinct viruses confers multi-virus resistance in transgenic plants through homology-dependent gene silencing. J Gen Virol 81:2103–2109

Kasschau KD, Carrington JC (1998) A counterdefensive strategy of plant viruses: suppression of posttranscriptional gene silencing. Cell 95:461–470

Lindbo JA, Dougherty WG (1992a) Pathogen-derived resistance to a potyvirus immune and resistant phenotypes in transgenic tobacco expressing altered forms of a potyvirus coat protein nucleotide sequence. Mol Plant Microbe Interact 5:144–153

Lindbo JA, Dougherty WG (1992b) Untranslatable transcripts of the tobacco etch virus coat protein gene sequence can interfere with tobacco etc h virus replication in transgenic plants and protoplasts. Virology 189:725–733

Lindbo JA, Silva-Rosales L, Proebsting WM, Dougherty WG (1993) Induction of a highly specific antiviral state in transgenic plants: Implications for regulation of gene expression and virus resistance. Plant Cell 5:1749–1759

Lius S, Manshardt RM, Fitch MMM, Slightom JL, Sanford JC, Gonsalves D (1997) Pathogen-derived resistance provides papaya with effective protection against papaya ringspot virus. Mol Breed 3:161–168

Manshardt RM (1992) Papaya. In: Hammerschlag FA, Litz RE (eds) Biotechnology of Perennial Fruit Crops. 21:489–511. CAB International, Wallingford, 550 pp

Manshardt RM (1998) ‘UH Rainbow’ papaya. Univ Hawaii Coll Trop Agric Human Resources. Germplasm G-1:2

Mau RFL, Gonsalves D, Bautista R (1989) Use of cross protection to control papaya ringspot virus at Waianae. Proceedings of the 25th Annual Papaya Industry Association Conference, 29–30 September, Hilo, HI, pp 77–84

Pang S-Z, Jan F-J, Gonsalves D (1997) Nontarget DNA sequences reduce the transgene length necessary for RNA-mediated topsovirus resistance in transgenic plants. Proc National Acad Sci USA 94:8261–8266

Powell-Abel P, Nelson RS, De B, Hoffmann N, Rogers SG, Fraley RT, Beachy RN (1986) Delay of disease development in transgenic plants that express the tobacco mosaic virus coat protein gene. Science 232:738–743

Purcifull D, Edwardson J, Hiebert E, Gonsalves D (1984) Papaya ringspot virus. CMI/AAB Descriptions of Plant Viruses. 292[84 Revised, July 1984]:8

Sanford JC, Johnston SA (1985) The concept of parasite-derived resistance – deriving resistance genes from the parasite’s own genome. J Theor Biol 113:395–405

Sanford JC, Smith FD, Russell JA (1992) Optimizing the biolistic process for different biological applications. Methods Enzymol 217:483–509

Tennant PF (1996) Evaluation of the resistance of coat protein transgenic papaya against papaya ringspot virus isolates and development of transgenic papaya for Jamaica. In: Department of Plant Pathology. Cornell University, Ithaca, pp 317

Tennant PF, Gonsalves C, Ling KS, Fitch M, Manshardt R, Slightom JL, Gonsalves D (1994) Differential protection against papaya ringspot virus isolates in coat protein gene transgenic papaya and classically cross-protected papaya. Phytopathology 84:1359–1366

Wang C-H, Yeh D-D (1997) Divergence and conservation of the genomic RNAs of Taiwan and Hawaii strains of papaya ringspot potyvirus. Arch Virol 142:271–285

Yeh S-D, Gonsalves D (1994) Practices and perspective of control of papaya ringspot virus by cross protection. In: Harris KF (ed) Advances in Disease Vector Research. Springer-Verlag, New York, pp 237–257

Yeh SD, Gonsalves D (1984) Evaluation of induced mutants of papaya ringspot virus for control by cross protection. Phytopathology 74:1086–1091

Fc Receptor Homologs (FcRH1–5) Extend the Fc Receptor Family

R.S. DAVIS[1], G. DENNIS JR.[2,*], H. KUBAGAWA[3], and M.D. COOPER[4]

1 Introduction

Receptors for the Fc portion of immunoglobulins (FcR) are expressed broadly among cells of the immune system (RAVETCH and KINET 1991; DAERON 1997). The differential expression of these Ig isotope-specific receptors enable them to modulate cellular and humoral immunity by

[1] Division of Hematology/Oncology and Department of Medicine, University of Alabama at Birmingham, AL 35294-3300, USA
[2] Department of Microbiology, University of Alabama at Birmingham, AL 35294-3300, USA
[3] Department of Pathology, University of Alabama at Birmingham, AL 35294-3300, USA
[4] Howard Hughes Medical Institute, University of Alabama at Birmingham, 378 Wallace Tumor Institute, Birmingham, AL 35294, USA
* *Present address*: Laboratory of Immunopathogenesis and Bioinformatics, Clinical Services Program, National Institute of Allergy and Infectious Disease, Science Applications International Corporation-Frederick, MD, USA

RSD and GD Jr. contributed equally to this analysis

linking their antibody ligands with effector cells. These cellular receptors have the ability to sense humoral concentrations of antibody, initiate cellular responses in host defense, and participate in autoimmune disorders (RAVETCH and BOLLAND 2001). The diverse regulatory roles of FcR depend upon their Ig isotype specificity, cellular distribution, and cytoplasmic signaling elements. The different FcR molecules share similarities in their ligand-binding subunits and overall extracellular structures, but they differ in their transmembrane and cytoplasmic regions in two major ways. These receptors either have self-contained inhibitory or activating signaling motifs or instead they use charged residues in their transmembrane region to pair with signal-transducing adaptor chains possessing activating signaling motifs.

Our view of the FcR family has been significantly expanded over the last 5 years by the characterization of the FcαR homologs in mice, paired Ig-like receptors (PIR), and their closest relatives in humans, the Ig-like transcripts/leukocyte Ig-like receptors (ILT/LIR) (BORGES et al. 1997; HAYAMI et al. 1997; KUBAGAWA et al. 1997; SAMARIDIS and COLONNA 1997). This multigene family in humans includes the FcαR (CD89), natural killer cell Ig-like receptors (KIR/CD158), and ILT/LIR (CD85), all of which are located in a chromosome 19 region known as the leukocyte receptor cluster (LRC) (KREMER et al. 1992; WAGTMANN et al. 1997; WENDE et al. 1999; WILSON et al. 2000; BARTEN et al. 2001). These Ig-like families belong to a larger group of receptors characterized by their possession of common cytoplasmic tyrosine-based signaling motifs. There are two different signaling subclasses which have either immunoreceptor tyrosine-based activation motifs (ITAM) containing two repeats of the consensus sequence Y-X-X-L/I spaced by 6–8 amino acids (E/D)-X-X-Y-X-X-(L/I)-X_{6-8}-Y-X-X-(L/I), or immunoreceptor tyrosine-based inhibitory motifs (ITIM) with a six amino acid consensus sequence (I/V/L/S)-X-Y-X-X-(L/V) (RETH 1992; DAERON et al. 1995; VELY and VIVIER 1997; GERGERLY et al. 1999; RAVETCH and LANIER 2000). Following ligand binding of the activating receptor complexes, tyrosines in the ITAM are rapidly phosphorylated by src family kinases to initiate a cascade of signaling events that trigger cellular activation. In the case of the ITIM-bearing receptors, the phosphorylated tyrosines provide a docking site for phosphatases containing SH-2 domains which can abrogate cellular activation via signaling pathways that depend upon tyrosine phosphorylation (UNKLESS and JIN 1997; LONG 1999). The balance between the activating and inhibitory receptor pairs can modulate cellular responses to a variety of stimuli.

The genes that encode the earliest defined FcRs, namely *FcγRI*, *FcγRII*, *FcγRIII*, and *FcεRI*, are located on the long arm of human chromosome 1 in a 1q21–23 locus (Fig. 1) that lies centromeric to the polymeric Ig receptor (*pIgR*) and *Fcα/μR* genes (1q32) (Tepler et al. 1989; Qui et al. 1990; Krajci et al. 1991; Oakley et al. 1992; Shibuya et al. 2000). These FcRs on chromosome 1 share limited sequence similarity with *FcαR* and its relatives in the LRC on chromosome 19 (Dennis et al. 2000), and they share common signaling components. The FcγRI, FcγRIIIA, FcεRI, and FcαR are all activating types of receptors that share an ITAM-containing γ-chain in common (FcRγc) (Perez-Montfort et al. 1983; Hibbs et al. 1989; Ernst et al. 1993; Scholl and Geha 1993; Pfefferkorn and Yeaman 1994; Morton et al. 1995).

The concentration of more than 24 genes in the 19q13 LRC locus that share a high degree of homology in their respective extracellular domains suggested by analogy that a similar cluster of *FcγR* and *FcεR* gene relatives might exist in the chromosome 1q21–23 region. Therefore, in order to search for new members of this FcR family, we identified a consensus amino acid motif through a comparison of the FcγRI (CD64), FcγRII (CD32), FcγRIII (CD16), and pIgR extracellular regions. When this consensus motif was used in a GenBank protein database query, genomic clones were identified that were found to contain *FcR* relatives that we have provisionally named the Fc receptor homolog (*FcRH*) family (Davis et al. 2001). These genes have also been identified independently as immunoglobulin superfamily receptor translocation associated genes (*IRTA*) through an analysis of the breakpoints of a t(1;14)(q21;q32) chromosomal translocation from a multiple myeloma cell line (Hatzivassiliou et al. 2001). The two different strategies used to identify these genes provide clues to their likely biologic roles as FcR-related molecules with diverse signaling characteristics and oncogenic potential. This review focuses on the prospective contributions of the newly characterized *FcRH/IRTA* multigene family to the biologic, structural, and phylogenetic understanding of the greater Fc receptor family.

2 The FcR Homologs

A novel family of genes that are related to previously identified *FcR* genes was independently discovered by two research groups. With a focus on defining commonly observed rearrangements at the 1q21 locus in B lymphoid malignancies, the Dalla-Favera group has characterized a t(1;14)(q21;q32) chromosomal translocation in a multiple myeloma cell line (Hatzivassiliou et al. 2001). Sequence analysis of the cloned breakpoint defined a reciprocal translocation involving a novel transmembrane immunoglobulin superfamily (IgSF) member, the immunoglobulin superfamily receptor translocation associated gene 1 (*IRTA1*), juxtaposed with the intron between the third constant region (*CH3*) and the transmembrane domain exons of IgA_1 (α_1). The cloning of the *IRTA1* gene and further analysis of the locus surrounding the breakpoint led to the discovery of a second gene, *IRTA2*, immediately centromeric to *IRTA1*.

With the recent recognition of a very large number of *FcR* relatives in the 19q13 LRC, we hypothesized that a similar extended family might exist in the chromosome 1q region where the other classic *FcRs* are known to reside. In our search for FcR-related family members, we used a consensus sequence to query the National Center for Biotechnology Information (NCBI) GenBank database. This 32 amino acid sequence corresponded with the amino terminal sequences of the second Ig domains of the FcγRs and the third Ig domain of pIgR. When this consensus motif was employed using the basic local alignment search tool (BLAST) algorithm in the NCBI protein database, we identified several overlapping bacterial artificial chromosomes (BAC) that mapped to the 1q21–22 region (ALTSCHUL et al. 1990). These BAC clones contained multiple genes that encoded putative

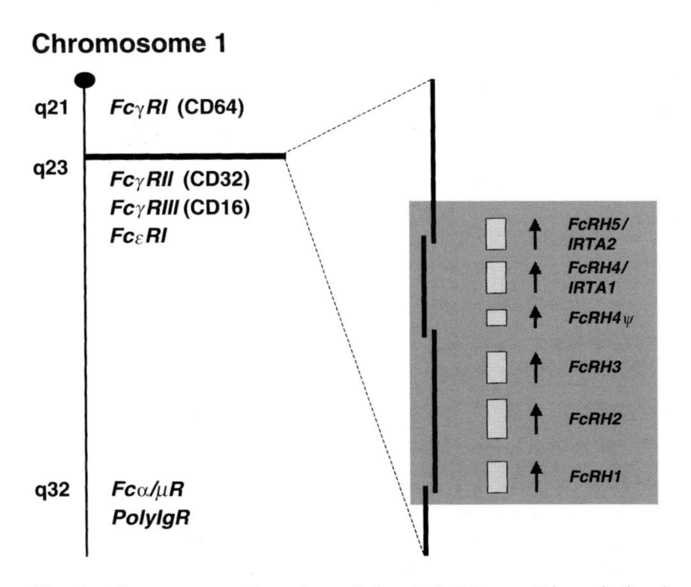

Fig. 1. Chromosomal location of the *FcRH* locus. The relative location of *FcRH* and *FcR* genes on chromosome 1 is approximated by their position according to the GenBank Mapview database. The *FcRH* locus is shown in relation to the other FcR-related genes in the region. The *FcRH* genes, next to the BAC clones that span the locus, are transcriptionally oriented towards the centromere

IgSF members we provisionally call the Fc receptor homologs (FcRH) (Davis et al. 2001) (Fig. 1).

Isolation of the FcRH cDNAs yielded five unique, but related sequences. FcRHs one through five encode type I transmembrane proteins with full-length sequences of 429 amino acids, 508 amino acids, 734 amino acids, 515 amino acids, and 977 amino acids, respectively (Fig. 2). For two of the FcRH genes, potential splice variants that modify the open reading frame (ORF) have been characterized. Transcripts for two potential secreted variants and a glycosylphosphatidylinositol (GPI)-anchored form have been identified for *IRTA2* (*FcRH5*). A truncated extracellular transmembrane form of *FcRH2* may also exist (Xu et al. 2001). All five of the predicted FcRH proteins possess similar hydrophobic signal peptides, three to nine Ig-like domains with three to eight potential N-linked glycosylation sites, transmembrane segments of 22–23 amino acids, and cytoplasmic tails

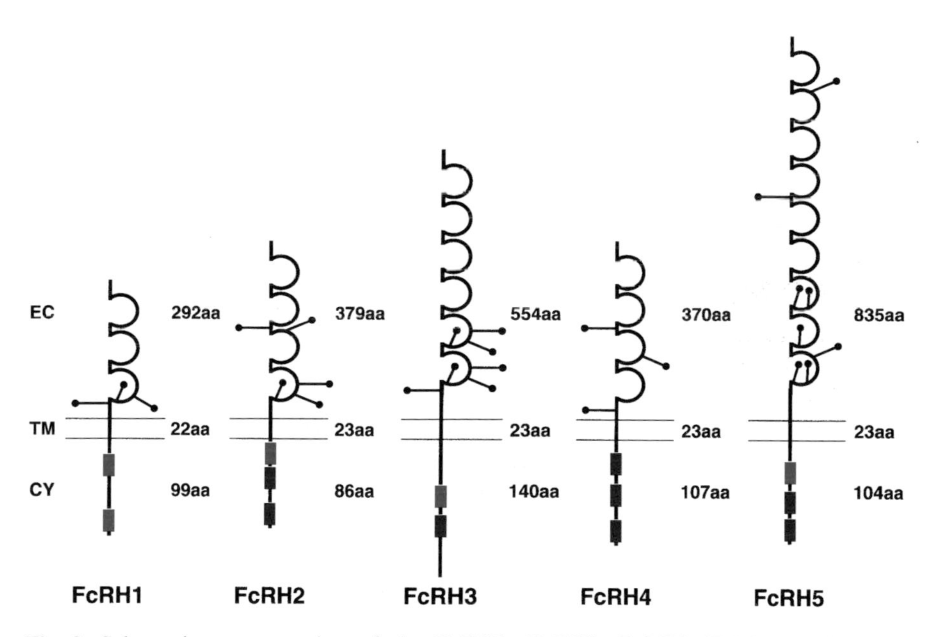

Fig. 2. Schematic representation of the FcRH1, FcRH2, FcRH3, FcRH4, and FcRH5 structural diversity. The five cDNAs are predicted to encode type I transmembrane proteins with similar extracellular domains, but different cytoplasmic regions. The predicted mature proteins contain extracellular (*EC*) regions with variable numbers of Ig-like domains and multiple sites for N-linked glycosylation (•—). Transmembrane (*TM*) domains are indicated. The cytoplasmic (*CY*) region of FcRH1 contains two potential ITAMs (*green boxes*). FcRH2 has one consensus ITAM and two ITIM (*red boxes*) consensus sequences. FcRH3 has the longest cytoplasmic tail with one ITAM and one ITIM. FcRH4 has three potential ITIMs and FcRH5 has one ITAM and two ITIMs. Amino acid (*aa*) lengths of each region are indicated

of 86–140 amino acids that contain consensus sequences for ITAM, ITIM, or both.

A remarkable diversity in the signaling potentials of the individual FcRH/IRTA is indicated by variations in makeup of the different FcRH cytoplasmic regions. FcRH1 has a long cytoplasmic tail of 99 amino acids that contains consensus sequences for two potential ITAM and a "hemi-ITAM" sequence, so called because it contains only one consensus tyrosine. Interestingly, a similar motif is evident in the rat and mouse PIR-B and in the human ILT-1/LIR-2 molecules. The cytoplasmic tail of FcRH2 is shorter and contains one ITAM and two ITIM consensus motifs separated by 22 amino acids. FcRH3 has the longest cytoplasmic tail, which contains one potential ITAM, one ITIM and a hemi-ITAM sequence near the carboxy terminus. FcRH4/IRTA1 has three consensus ITIM sequences in its cytoplasmic tail of 107 amino acids. The last member, FcRH5/IRTA2, has a cytoplasmic tail of 104 amino acids that contains one potential ITAM followed by two ITIM separated by 24 amino acids. Biochemical studies of SPAP-1 (FcRH2) in cotransfection experiments have demonstrated SHP-1 recruitment potential, but failed to show SHP-2 recruitment (Xu et al. 2001). The analysis of the signaling properties of the FcRH family members is thus just beginning. Future studies will be needed to delineate the signaling pathways engaged by the different FcRHs, and to determine whether the utilization of these pathways may vary in the different cell types that have been found to express the *FcRH*.

3 Cellular Distribution of FcRH Expression

The tissue distribution for *FcRH1*, *FcRH2*, *FcRH3*, *FcRH4/IRTA1* and *FcRH5/IRTA2* expression has been surveyed by Northern blot analysis using gene-specific probes (Davis et al. 2001; Hatzivassiliou et al. 2001). All five of the *FcRH/IRTA*-specific probes hybridize with spleen and lymph node transcripts, although *FcRH4/IRTA1* expression was detected at relatively low levels in these tissues. Another common feature of the family members is suggested by the identification of multiple sized transcripts in the lymphoid tissues, a finding that may reflect alternative splicing or differential polyadenylation. *FcRH/IRTA* transcripts were most abundant in peripheral lymphoid tissue samples, but they could also be found in other non-lymphoid tissues with the single exception of *FcRH4/IRTA1* transcripts. *FcRH1* appears to be the most broadly expressed since its transcripts were identified in heart, skeletal muscle, kidney, liver, and, in less abundance, placenta. *FcRH2* transcripts were also tentatively identified in the kidney. *FcRH3* expression was seen in bone marrow and peripheral blood cells, albeit in much lower levels than in lymph node and spleen. *FcRH3* transcripts of variable sizes were also present at lower levels in kidney, liver, placenta, and lung. *FcRH4/IRTA1* was detected only in the spleen and lymph node cells indicating a lymphoid tissue-restricted pattern of expression for this FcRH family member. In addition to the peripheral lymphoid tissues, *FcRH5/IRTA2* was also found to be expressed in bone marrow. A relatively broad expression pattern of this FcRH family member is suggested by the presence of its transcripts in small intestine, skeletal muscle, and heart. Thus, while all of the *FcRH* family members are expressed in secondary lymphoid tissues, their possible expression in a variety of non-lymphoid tissues suggests that they could play a relatively broad physiological role or, more likely, multiple roles.

Expression of the FcRH family has been evaluated by reverse transcriptase polymerase chain reaction (RT-PCR) analysis in a variety of cell lines representing different hematopoietic lineages (Davis et al. 2001). Expression of all five *FcRH* family members could be detected in almost every mature B cell line tested (Table 1). *FcRH2*, *FcRH3*, and *FcRH4* expression was not seen in the other types of hematopoietic cells that were tested. *FcRH1* and *FcRH5* expression was uniquely identified in pro-B cell lines. *FcRH1* expression was also seen in myeloid and T cell lines, but not in an erythroid cell line.

The analysis of sorted subpopulations of peripheral blood cells indicated that *FcRH1*, *FcRH2*, *FcRH3*, and *FcRH5* are expressed at relatively

Table 1. FcRH transcript expression in human cell lines

Cell type	Cell line	*FcRH1*	*FcRH2*	*FcRH3*	*FcRH4*	*FcRH5*
Pro B	REH	+	–	–	–	+
	Nalm16	+	–	–	–	+
Pre B	697	–	–	–	–	–
	OB5	–	–	–	–	–
	207	–	–	–	–	–
B	Ramos	+	+	+	–	+
	Daudi	+	+	+	+	+
	Raji	+	+	+	+	+
T	Jurkat	+	–	–	–	–
Monocytic	THP-1	+	–	–	–	–
Myelomonocytic	U937	+	–	–	–	–
Promyelocytic	HL-60	+	–	–	–	–
Myelocytic	KG-1	+	–	–	–	–
Erythroid	K562	–	–	–	–	–

FcRH transcript expression in human hematopoietic cells. Expression (+) or lack of expression (–) is based on reverse transcriptase-PCR analysis using FcRH specific primers.

high levels in primary B cells, whereas *FcRH4* was expressed at only trace levels. *FcRH3* expression was observed in circulating T cells, in which trace levels of *FcRH1* transcripts were also found. *FcRH1* expression was also evident in the circulating granulocytes. Analysis of *FcRH* expression in B lineage cell subsets from tonsillar samples revealed that these genes are differentially expressed by mature B lineage cells (Davis et al. 2001). The subdivision of tonsillar B cells on the basis of their differential cell surface expression of IgD and CD38 has been shown to correlate with the progressive stages of B cell maturation: follicular mantle B ($IgD^{+}CD38^{-}$), pre-germinal center ($IgD^{+}CD38^{+}$), germinal center ($IgD^{-}CD38^{+}$), memory B ($IgD^{-}CD38^{-}$), and plasma cells ($CD38^{++}$) (Pascual et al. 1994). RT-PCR analysis of these B cell subpopulations indicated coordinate expression of all five *FcRH* transcripts in follicular mantle B cells and memory B cell subpopulations, whereas none of the *FcRH* transcripts were detected in the pre-germinal center subset of B cells. More distinctive expression patterns were identified for the different *FcRH* family members in other B cell subsets. *FcRH1*, *FcRH2* and *FcRH3* transcripts were found in germinal center B cells, whereas *FcRH5* expression uniquely extended into the mature plasma cell stage. In situ hybridization studies of the topographical distribution of B cells that express *FcRH4/IRTA1* and *FcRH5/IRTA2* have indicated *FcRH4/IRTA1* expression by the marginal zone B cells, whereas *FcRH5/IRTA2* expression was also evident in germinal center B cells and immunoblasts. This analysis indicates a highly regulated pattern of differential *FcRH* gene expression during B cell activation and differentiation into either memory B cells or antibody secreting plasma cells. The generation of FcRH-specific antibodies in the future will allow a more precise definition of the cellular distribution of the different FcRH family members.

4 The Human *FcRH* Locus

The *FcRH* locus spans a ~300-kb region within the *FcR* cluster on human chromosome 1. The *FcγRI* is located centromeric of the *FcRH* locus that maps cytogenetically in the 1q21–22 region (see Fig. 1). According to the GenBank Mapview database, the *FcγRII*, *FcγRIII*, and *FcεRI* genes, as well as the gene for *CD3ζ*, are located to the telomeric side of this region. The *FcRH* genes, including the *FcRH4* pseudogene, lie in the same transcriptional orientation toward the centromere. The overall genomic structure of the five functional genes is conserved among the family members, but they have different numbers of Ig-like domains (Fig. 3). In four of the *FcRH*

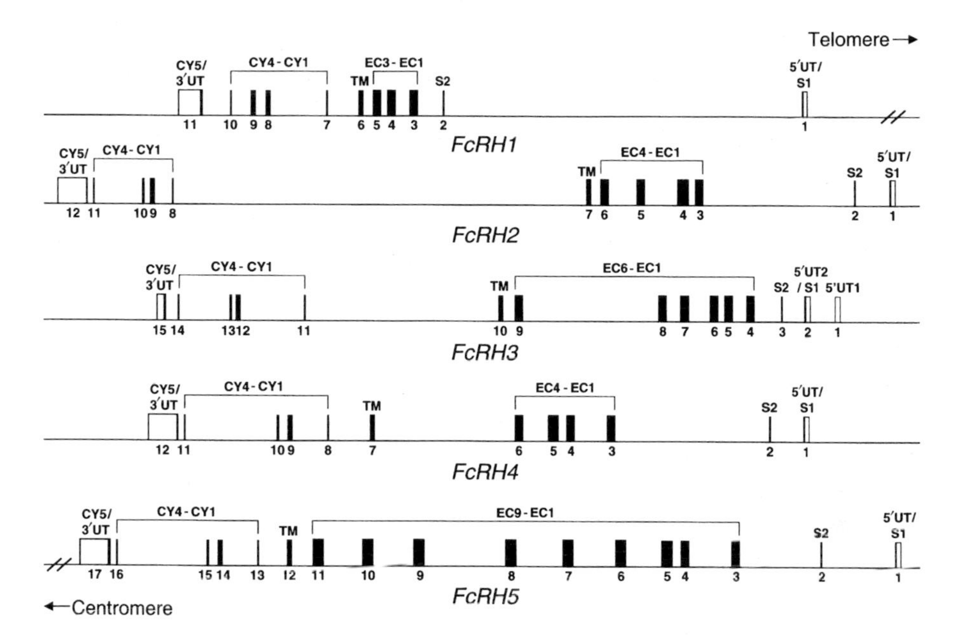

Fig. 3. Genomic organization of the *FcRH* family members. Genomic structure was determined by comparing cDNA clones with the BAC clones that overlap the region. Exon/intron boundaries were characterized by sequence comparisons and the AG/GT rule. Genes are depicted from most telomeric (*FcRH1*) to centromeric (*FcRH5*) and lie in their proper transcriptional orientation. Exons are *numbered* and indicated by *rectangles*: untranslated (*open*) and translated (*closed*). Domains are listed as: *UT*, untranslated; *S*, signal peptide; *EC*, extracellular; *TM*, transmembrane; and *CY*, cytoplasmic regions

family members the first exon (5′UT/S1) is composed of the 5′UT region, the ATG translation initiation site, and the beginning of a split signal peptide. The exception to this rule is *FcRH3* in which the first exon (5′UT1) contains the initial part of the 5′UT and the second exon (5′UT2/S1) contains the terminal portion of the 5′UT, the ATG translation initiation site, and the initial portion of a split signal peptide. The third exon of *FcRH3* and the second exon in the other *FcRH*s consists of 21bp and encodes the second half of the signal peptide. This "miniexon" is conserved in all *FcRH* family members and is also a consistent feature of the neighboring *FcγR* and *FcεR* genes (van de Winkel and Capel 1993; Pang et al. 1994). In contrast, this characteristic is not seen in the more distant gene relatives in the LRC region on chromosome 19. Instead, *FcαR* and the *KIR* and *ILT*/*LIR* family members contain a 36-bp S2 "miniexon" (de Wit et al. 1995; Selvakumar et al. 1997; Wilson et al. 1997; Torkar et al. 1998). This distinction is also evident in the murine members of these different branches of the ancestral FcR tree. The *PIR* genes, which lie on mouse chromosome 7 in a telomeric region syntenic with the human chromosome

19q13 LRC locus, also possess a 36-bp "miniexon", whereas the mouse *FcγR* and *FcεR* orthologs have a 21-bp S2 exon (Kulczycki et al. 1990; Ye et al. 1992; Alley et al. 1998). Each of the extracellular Ig-like regions of these receptor genes is encoded by a separate exon (EC), a feature that is characteristic of IgSF members (Williams and Barclay 1988). The exons encoding the transmembrane portions include as well the membrane proximal extracellular sequence (TM) and the beginning of the cytoplasmic region. The overall structure of the cytoplasmic exons is also conserved in that the cytoplasmic regions of all *FcRH* are encoded by five exons, CY1–CY5, the last of which contains the termination codon of the open reading frame and the beginning of the 3′UT region. All of the intron/exon junctions follow the GT-AG rule, except at a single boundary within the EC encoding region of the *FcRH2* gene. This difference entails a GT>GC change at the 5′ splice donor boundary of EC2 with the fourth intron. This cryptic site may thus give rise to alternatively spliced products for *FcRH2*.

Intron/exon boundaries of the 5′UT/S1, S2, EC, and TM of all *FcRH* family members follow the phase 1 splicing pattern, i.e., splicing occurs after the first nucleotide of the triplet codon. This is also true for the TM/CY1 junction, whereas the CY1/CY2 boundaries indicate phase 2 splicing. CY2/CY3 and CY3/CY4 are both in phase 1, and the last cytoplasmic exon/intron junction, CY4/CY5–3′UT, is phase 0. A previous analysis of available information on the genomic organization of IgSF receptors containing ITIM has noted that cytoplasmic exons encoding functional ITIMs are preceded by phase 0 intron/exon boundaries (Daeron and Vivier 1999). Correspondingly, the last cytoplasmic exon of *FcRH2* harbors one putative ITIM, as does *FcRH3*, *FcRH4*, and *FcRH5*.

The similarity between the genomic structure of the *FcRs* and the immediately adjacent *FcRH*s infers their derivation from a common ancestor through extensive gene duplication and diversification that is reflected in the numbers and complexity of EC exons and CY exons. A consistent feature among the FcR family members and their 1q21–23 *FcRH* relatives is the S2 21-bp "miniexon", although other structural modifications in these genes suggest subtle differences in their phylogenetic roots. FcγRI appears to be closest FcRH relative with respect to its EC1-EC3 exons. The FcγRI has an EC3 exon that encodes an Ig domain that is unique among the classical FcRs, but is shared in common with all of the FcRH family members. However, the cytoplasmic region of FcγRI lacks signaling tyrosine-based signaling elements, and thus must associate with FcRγc in order to be able to transduce intracellular signals. On the other hand, the three *FcγRII* genes have CY exons that encode either ITAM (*FcγRIIA,C*) or ITIM (*FcγRIIB*) consensus motifs in their gene specific isoforms (Daeron

1997). The FcRH, therefore, more closely resemble the *FcγRII* genes with regard to their cytoplasmic structure.

5 The Classical Fc Receptors

The defining feature of the previously recognized *FcR*s that surround the *FcRH* locus on 1q21–23 is their ability to bind immunoglobulin. These receptors have been extensively characterized and their protein products have been shown to play important roles in receptor-mediated regulation of the immune system. They may provide either positive or negative signals to modulate inflammatory responses mediated by cells of the immune system (HEYMAN 2000). Disruption of the modulatory effects mediated by the FcRs has been implicated in autoimmune disorders, such as systemic lupus erythematosus and rheumatoid arthritis, in allergic hypersensitivity, and in hematopoietic malignancies (SYLVESTRE and RAVETCH 1996; CLYNES et al. 1998; YUASA et al. 1999; CALLANAN et al. 2000; KLEINAU et al. 2000; NAKAMURA et al. 2000; RAVETCH and BOLLAND 2001). In addition, the FcRs may play crucial roles in anti-tumor immunity and in determining the therapeutic effects of monoclonal antibodies (DYALL et al. 1999; CLYNES et al. 2000). Cytogenetic studies have indicated that the human chromosome 1q21–23 region, where the *FcγRI*, *FcγRII*, *FcγRIII*, and *FcεRI* reside, is one of the most frequent sites for the occurrence of chromosomal aberrations in hematologic malignancies (OFFIT et al. 1991; CIGUDOSA et al. 1999). Analysis of the immunological capabilities of FcR-deficient mice indicates that the FcRs share an essential role in three types of inflammatory hypersensitivity reactions: immediate hypersensitivity, cytotoxic inflammatory responses, and immune complex-mediated inflammation (RAVETCH and BOLLAND 2001). On the other hand, the FcR-deficient mice have not been found to have obvious defects in innate immunity or in susceptibility to a variety of pathogens (RAVETCH and CLYNES 1998).

The currently defined IgE and IgG FcRs fall into two general classes: (1) activating receptor complexes that exert their signaling function through ITAMs, and (2) inhibitory receptors that are endowed with ITIMs (DAERON 1997). The ITAMs utilized by the activating receptors may be intrinsic to their cytoplasmic regions, as is the case for FcγRIIA and FcγRIIC, but they are acquired more commonly through association of the ligand-binding chains with adaptor transmembrane molecules that contain ITAM. The presence of a positively charged amino acid in the transmembrane region of the ligand-binding chains of FcεRI, FcγRI, and

FcγRIII endows these receptor units with the capacity to partner with the ITAM containing FcRγc to mediate cellular activation. On the other hand, the inhibitory FcR family member, FcγRIIB, has no positively charged amino acid in its transmembrane region, but instead inhibits cellular activation through a cytoplasmic ITIM that upon phosphorylation can serve as a docking site for protein tyrosine phosphatases. The remarkable diversity of the tyrosine-based motifs of the FcRH relatives implies that they may possess an even greater degree of complexity than the classical FcR in their signaling capabilities.

6 Conservation of FcRH and FcR Extracellular Domains

Pairwise comparisons of FcR and FcRH amino acid sequences suggest a significant degree of conservation in their extracellular regions. Comparative analysis of their individual Ig-like subunits reveals similarities that go well beyond simple separation into two distinct genetic lineages. An unrooted phylogenetic tree segregates a combined total of 37 domains from both families into five distinct clusters, three of which contain domains from both FcR and FcRH family members (Fig. 4). The clusters indicated by this analysis are color-coded in Figure 4 on the basis of the major branching points. The yellow cluster contains five FcRH domains, one from each FcRH family member along with one from FcγRI, namely domain three (D3), a domain that is unique among the FcRs, but one that is conserved in all of the FcRH family. The red cluster contains exclusively N-terminal domains, one from each of the FcRs, and the first domains of FcRH3, FcRH4, and FcRH5. The second domain (D2) of the classical FcRs is the major region of interaction with Ig, and this domain clusters with homologous domains in four of the five FcRH family members. The D1-D2 (red–blue) pattern of arrangement seen in the FcR family is conserved in their FcRH3, FcRH4, and FcRH5 relatives. The structural relationship between these two domains is critical for Fc binding by the classical FcRs, wherein D1 bends at an acute angle in order to expose the ligand-binding region of D2. Conservation of this domain-tandem in the majority of the FcRH members may thus provide an important clue to the nature of the FcRH ligands. This analysis reveals that the FcRH and FcR Ig-like domains are related in a manner that extends beyond subfamily boundaries, a property that would be expected to be shared by modern descendents of an ancient gene family. The data also indicate that the FcRHs are much more closely related to their *FcR* neighbors on chromo-

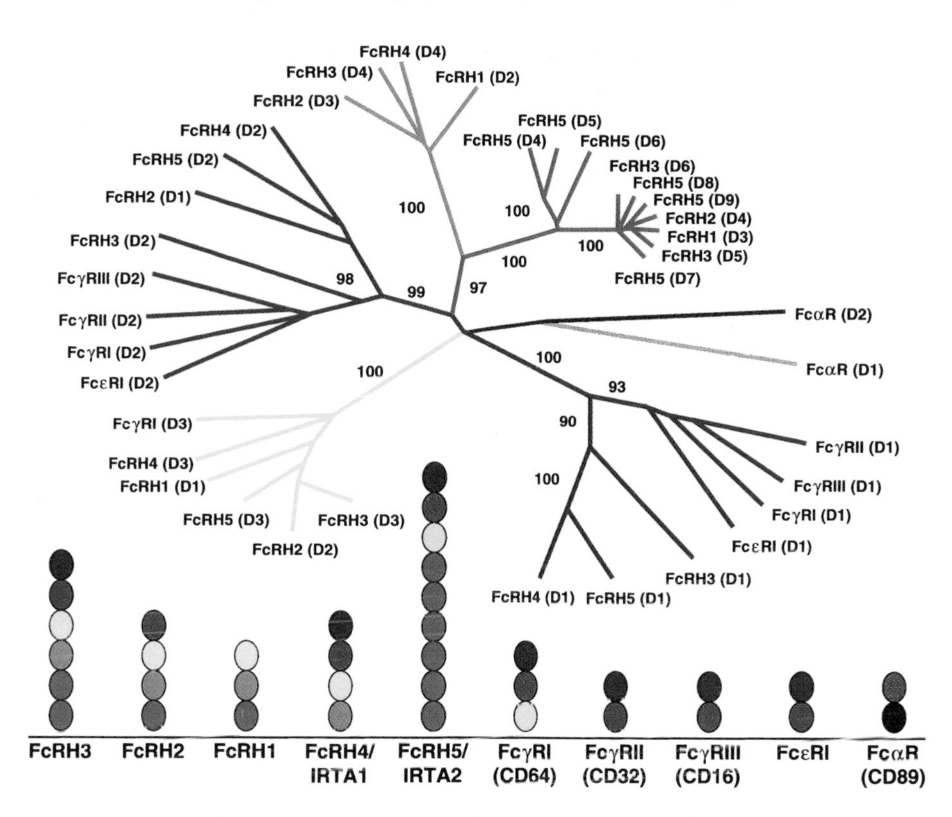

Fig. 4. Depiction of the extracellular Ig-like domains in FcRH and FcR and a phylogenetic tree inferring the relationships among them. Each domain is color-coded in a manner corresponding to its clustering position within the tree (see text). An all-against-all amino acid sequence alignment of the Ig-like domains was performed using ClustalX. The unrooted tree topology was estimated using TreeCon, where branch values represent percentage Bootstrap support after 500 replicates (Van de Peer and De Wachter 1993)

some 1 than to their FcαR relative on chromosome 19. The FcαR D1 and D2 domains can be seen to noticeably depart from the rest of the analyzed subunits in this analysis. Although similar Ig domain subunits are shared among the FcRH and FcR subfamilies, the individual FcRH receptors are composed of unique domain combinations. Not only are individual domains related in a predictable fashion, but these subunits illustrate conservation in the membrane-distal to membrane-proximal ordering of extracellular domain composition among family members. Divergences between the FcRH and FcR subfamily members are evidenced by the apparent duplication of novel membrane-proximal Ig-like domains (green). The collective evidence of chromosomal proximity, genomic structural, and Ig-like domain homology nevertheless clearly suggest that the *FcRH* genes are close phylogenetic relatives of the *FcR* family.

7 Paired Ig-Like Receptor Relatives

The IgA binding *FcαR* residing on human chromosome 19 shares limited sequence similarity with the *FcRs* and the *FcRH*s on chromosome 1. FcαR orthologs have not yet been identified in non-primates, whereas orthologs of IgE and IgG FcRs have been found to be broadly represented in mammals. Phylogenetic analysis of the mammalian FcRs predicts that the closely related FcαR in humans and the FcγR in cows represent a distinct class of mammalian FcRs. In this regard, the bovine *Fcγ2R* and human *FcαR* are encoded in syntenic regions of bovine chromosome 18 and human chromosome 19, where no other FcRs have been identified (Zhang et al. 1995).

The search for FcαR orthologs in mice led to the surprising discovery of a set of paired Ig-like receptors, *PIR-A* and *PIR-B* (Kubagawa et al. 1997; Hayami et al. 1997). Like its FcαR relative, and FcγRI, FcγRIII, and FcεRI as well, the PIR-A molecules form activating receptor complexes via their association with the FcRγc transmembrane co-adaptor molecule. The approximately eight PIR-A isoforms are encoded by separate genes, each of which possesses a relatively large extracellular region containing six Ig-like domains with limited sequence variations. The extracellular amino acid sequence of PIR-B shares greater than 90% identity with the extracellular sequences of the different PIR-A isoforms. The invariant PIR-B molecule is encoded by a single gene, possesses an uncharged transmembrane segment, and functions as an inhibitory receptor through the action of its ITIM-containing cytoplasmic region. The activating and inhibitory signaling potentials of PIR-A and PIR-B, coupled with their conserved Ig-like EC domains, are highly reminiscent of the IgG and IgE FcR family members, although neither PIR-A nor PIR-B has been shown to bind immunoglobulins. Indirect clues instead suggest that PIR-B may bind major histocompatibility complex (MHC) class I-like molecules (Ho et al. 1999).

The *PIR* gene family resides in a telomeric region of mouse chromosome 7 that is syntenic with the human chromosome 19q13.4 region. As indicated earlier, analysis of the region centromeric to the human *FcαR* gene has lead to the identification of a large monophyletic family of approximately 24 leukocyte receptor genes that are collectively termed the leukocyte receptor complex (LRC) of genes (Kremer et al. 1992; Wagtmann et al. 1997; Wende et al. 1999; Wilson et al. 2000). The products for many of these are expressed as activating and inhibitory receptor pairs. Included within the LRC are the *ILT*/*LIR* genes the closest human relatives of *PIR*, and the multigene *KIR* family.

The activating and inhibitory receptors comprising the LRC include the FcαR, ILT/LIR, KIR, leukocyte associated-inhibitory receptors (LAIR), and NKp46. It has been recognized previously that the Ig-like domains of the FcR and LRC family members share structural similarity in spite of significant divergence in their amino acid sequences (Dennis et al. 2000). However, the distinct ligand-binding specificities and lack of chromosomal synteny has made it difficult to understand the relationship between the LRC and FcR families. With the discovery of the FcR homologs and their chromosomal location amongst the *FcR* genes, the evolutionary relationships between the multigene families of activating and inhibitory receptors are becoming even more apparent. Our phylogenetic analysis of representatives from the FcRH, FcR, and LRC families depicts a robust relationship between these branches of the Ig gene superfamily (Fig. 5). This analysis reaffirms the close relationship between the FcRH and FcR. Mouse PIR is seen as the closest rodent relative of human LIR, while the chicken Ig-like receptor (CHIR) is seen as a distant relative that lies between the nodes containing LRC and FcRH/FcR family members (Dennis et al. 2000). Included in the analysis are receptors that have similar function, pIgR, Fcα/μR, and FcRn, and/or similar secondary structure, pIgR, Fcα/μR, and signal regulatory protein (SIRP). The segregation of these receptors outside of the common node shared by the FcRH, FcR and LRC families further suggests that the phylogenetic relationships between FcRH, FcR and LRC involve an ancient gene family that has gone through multiple rounds of duplication and diversification events. Another potential clue comes from related studies that have sought to trace the evolutionary history of the MHC genes where data based on models of whole genome duplication suggest that 1q21 and 19q13 are paralogous regions of the human genome (Kasahara 1999).

ILT-2/LIR-1 and their KIR relatives have been shown to recognize classical and non-classical MHC class I antigens (Moretta et al. 1996; Borges et al. 1997; Colonna et al. 1997; Cosman et al. 1997; Fanger et al. 1998; Lanier 1998a,b; Chapman et al. 1999). Notably, while the KIRs bind to MHC class I sites near the polymorphic α helices and peptide-containing grooves in the first domain of the MHC class I α chain, the ILT/LIR bind the non-polymorphic sites within the third domain of the MHC class I α chain (Fan et al. 1997; Chapman et al. 2000). The functional significance of the distinct recognition strategies used by KIR versus ILT/LIR remains unclear. What seems most clear is that the pairing of activating and inhibitory receptors with similar ligand-binding regions has been adopted as an important regulatory strategy in a variety of biological settings, the full extent of which has yet to be elucidated.

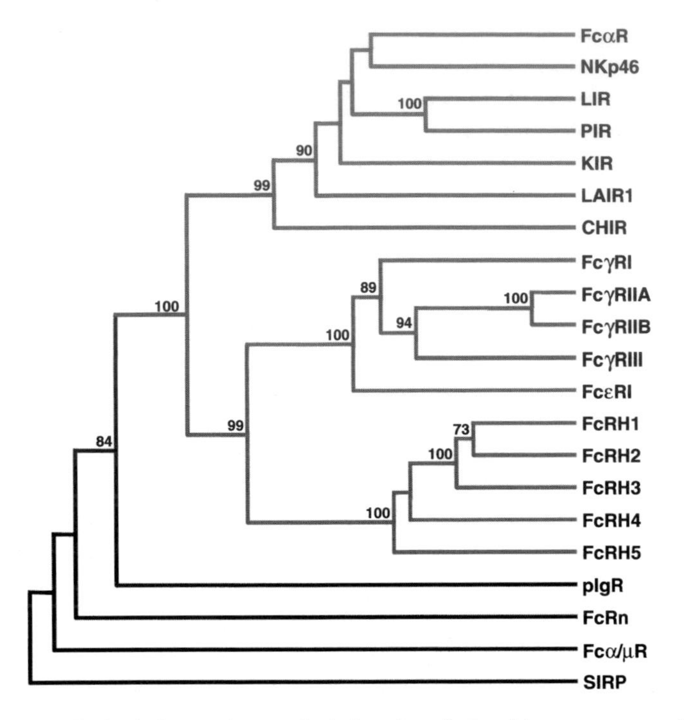

Fig. 5. A phylogenetic tree depicting the relationships among members of the greater FcR/PIR family. The extracellular amino acids of each receptor were aligned using ClustalX and the tree topology was estimated using TreeCon where branch values represent percentage Bootstrap support after 500 replicates. The node containing the greater FcR/PIR family shows robust branch support and is highlighted in *red*. Branches of PIR relatives are highlighted in *blue* and FcR relatives in *green*. Receptors that share similar function (Fc-binding), pIgR, FcRn, and Fcα/μR and receptors with similar secondary structure (Ig-like domains) pIgR, Fcα/μR, and SIRP were included in the analysis as a measure of tree topology. Segregation of these receptors outside of the greater FcR/PIR node supports the inferred phylogenetic relationships in the greater FcR/PIR family

8 Conserved FcR Structure Features

The structures for the extracellular Ig-like domains of FcγRIIA, FcγRIIB, FcγRIII, and FcεRI have been solved by X-ray crystallographic study (Garman et al. 1998; Maxwell et al. 1999; Sondermann et al. 1999; Sondermann et al. 2000) (Fig. 6A). Previously recognized as closely related receptors that share significant sequence similarity and biological function, the solution of their three-dimensional structures has led to an appreciation that tandem Ig-like domains are arranged at an acute angle to form a V-shaped structure. The individual Ig-like domains are characterized by an identical overall fold topology with five-stranded and three-stranded β-sheets facing each other to form a sandwich-like structure (Fig. 6A). This fold topology combines features of both I-type and C2-type Ig domains in a manner that has only been observed in the FcR family members and members of related gene families that reside on human chromosome 19q13 (SONDERMANN et al. 2001).

Mutagenesis studies of the *FcR* genes have indicated that the amino acid residues involved in Ig binding are located in the loop regions of the second extracellular domain (D2) and in the linker to the N-terminal domain (D1) (HULETT et al. 1995; TAMM and SCHMIDT 1997). The bent structure of the FcRs thus exposes the top of D2 and the linker in a manner that leads to the formation of the interaction site with immunoglobulins. The more recent solution of structures for the IgE-FcεRI and IgG-FcγRIII complexes confirm that Ig indeed interacts with these exposed D2 and linker regions (GARMAN et al. 2000; SONDERMANN et al. 2000). Unlike the FcεRI, FcγRIII rearranges upon binding with its IgG ligand to open the angle between D1 and D2 by approximately 10°, while the structure of FcεRI remains virtually unchanged in its bound-versus-free forms. Despite these minor differences, the modes of interaction observed in the two complexes are very similar and involve structural motifs that are highly conserved among IgE and IgG FcRs (Fig. 6B). The FcR consensus motif that was used in our identification of the FcR homologs contains highly conserved residues in the ligand-binding sites of the FcRs, and many of these conserved residues have been shown to interact directly with the Fc region of their immunoglobulin ligands (Fig. 6C). Interestingly, dissection of the ligand-binding region of FcαR by mutagenesis and domain swapping experiments has indicated that the D1 F/G loop of the FcαR is essential for IgA binding, thereby predicting a unique mode of interaction between the FcαR and IgA antibodies (MORTON et al. 1999; PLEASS et al. 1999).

9 Looking Ahead to the Potential Biologic Roles for the FcRH Family

The genomic location, amino acid sequence, and protein structure of the FcRH family members provide provocative clues to their functional potential. Currently available experimental techniques, such as X-ray crystallography, phylogenetic structure/sequence analysis, and sequence database biotechnology provide valuable theoretical and analytical tools that may yield additional insight into the function of the newly characterized

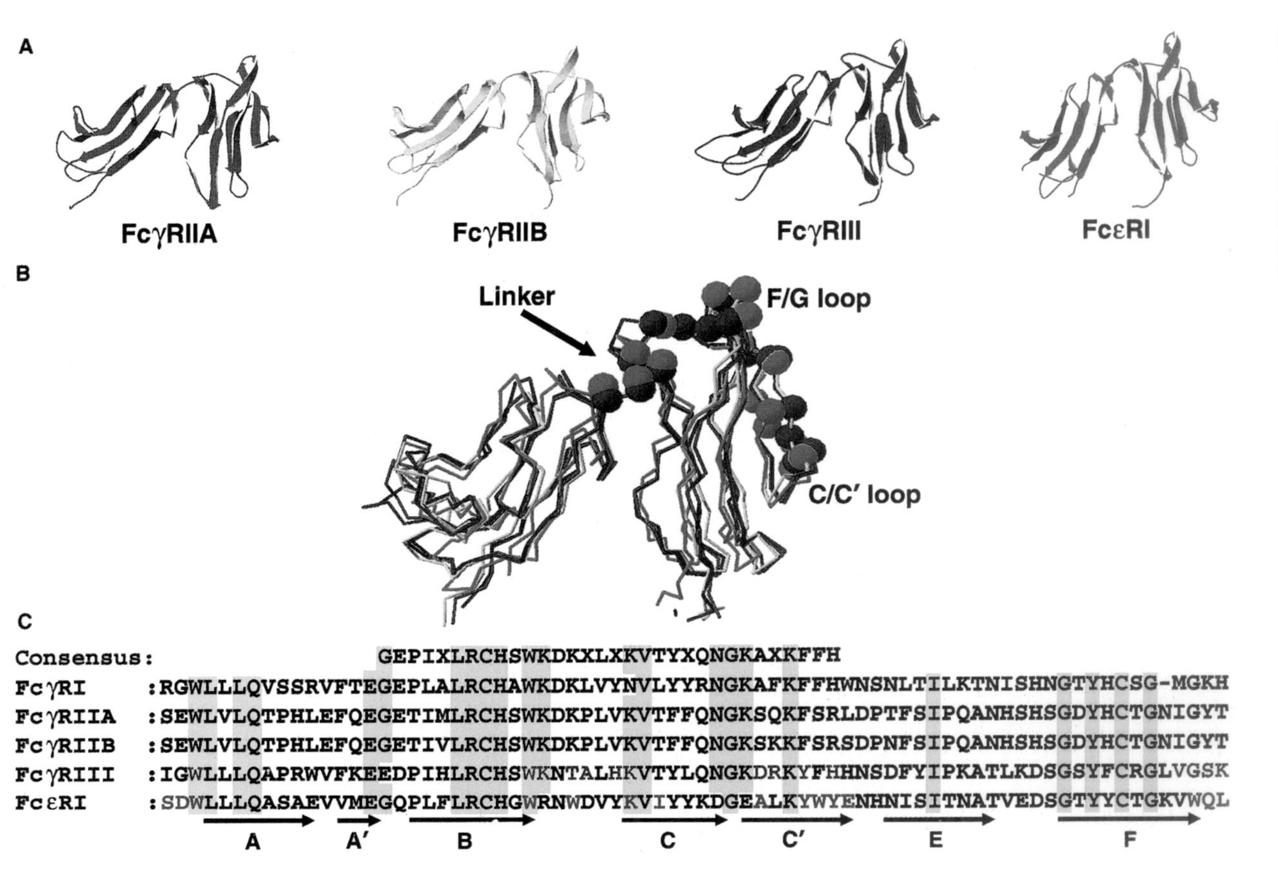
A
FcγRIIA
FcγRIIB
FcγRIII
FcεRI
B
Linker
F/G loop
C/C′ loop
C
Consensus: GEPIXLRCHSWKDKXLXKVTYXQNGKAXKFFH
FcγRI :RGWLLLQVSSRVFTEGEPLALRCHAWKDKLVYNVLYYRNGKAFKFFHWNSNLTILKTNISHNGTYHCSG-MGKH
FcγRIIA :SEWLVLQTPHLEFQEGETIMLRCHSWKDKPLVKVTFFQNGKSQKFSRLDPTFSIPQANHSHSGDYHCTGNIGYT
FcγRIIB :SEWLVLQTPHLEFQEGETIVLRCHSWKDKPLVKVTFFQNGKSKKFSRSDPNFSIPQANHSHSGDYHCTGNIGYT
FcγRIII :IGWLLLQAPRWVFKEEDPIHLRCHSWKNTALHKVTYLQNGKDRKYFHHNSDFYIPKATLKDSGSYFCRGLVGSK
FcεRI :SDWLLLQASAEVVMEGQPLFLRCHGWRNWDVYKVIYYKDGEALKYWYENHNISITNATVEDSGTYYCTGKVWQL
A
A′
B
C
C′
E
F

Fig. 6. Structural basis for FcR binding of antibody Fc regions. **A** Ribbon diagrams representing FcγRIIA (*blue*), FcγRIIB (*gold*), FcγRIII (*red*), and FcεRI (*green*) depicting domain 1 (D1) bent at an acute angle to the left relative to domain 2 (D2). **B** Superposition of FcR structures. The color and orientation of the structures are the same as above. Amino acid residues of FcγRIII and FcεRI that contact IgG1 and IgE are represented as *red* and *green space filled spheres*, respectively. **C** Amino acid sequence alignment of the ligand-binding regions for FcγRI-III and FcεRI. Conserved residues are highlighted in *yellow* and amino acids contacting Fc in FcγRIII and FcεRI are highlighted in *red* and *green*, respectively. The *bold arrows* indicate regions of β-sheets corresponding to FcγRIII. The consensus sequence employed in the discovery of the Fc homologs is shown atop the alignment spanning the ligand-binding region. Structural decorations and superpositions were created using Swiss-pdb Viewer (Guex and Peitsch 1997). The amino acid sequence alignment was produced with ClustalX (Thompson et al. 1997) and decorated with GeneDoc (Nicholas et al. 1997)

FcRH multigene family. For example, a structure based alignment of the consensus 32 amino acid sequence used to identify the *FcRH* family in the human genome reveals that this sequence spans the binding cleft of the Fc-binding region of the classical Fc receptors from which it was generated (Fig. 6C). The X-ray diffraction analysis of these structures indicates that the sequence and structural features of this region of the FcR family members are highly conserved. This constellation of findings strongly suggests that the three FcRH that possess this motif, FcRH3, FcRH4, and FcRH5, will have Ig Fc binding properties. Indeed, unpublished observations of the Della-Favera group support the conclusion that at least one FcRH family member, IRTA2/FcRH5, can bind IgG_1 and IgG_2 (Hatzivassiliou et al. 2001).

To date, FcγRIIB, an ITIM-bearing inhibitory receptor, and the Fcα/μR are the only classical FcRs that have been found on B lymphocytes (Amigorena et al. 1989; Shibuya et al. 2000). It has been widely held that the presence of the FcγIIB receptor on B cells acts to establish antibody feedback that can terminate the humoral response and lead to peripheral tolerance by binding the Fc portions of soluble immunoglobulins (Henry and Jerne 1968; Heyman 1999; Bolland and Ravetch 2000). This mechanism involves the simultaneous interaction of antigen-antibody complexes with the FcγRIIB and the B cell receptor (BCR) for antigen. When these receptors are co-engaged, the ITIM of FcγRIIB is tyrosine phosphorylated, thereby recruiting src homology phosphatases, which in turn inhibit BCR/ITAM mediated cellular activation and proliferation (Daeron et al. 1995; D'Ambrosio et al. 1995; Ono et al. 1996). The Fcα/μ receptor has not been shown to participate directly in B cell activation, but may function to promote phagocytosis of antigen-IgM antibody complexes (A. Shibuya, personal communication). Notably, none of the activating members of the classical FcRs have been found on B cells.

The finding of a family of five previously unrecognized receptors with diverse signaling potential and Fc receptor homology that are preferentially expressed on B cells has many potential implications. Given that these genes can be expressed in a highly coordinate, differential fashion by B cells that are undergoing proliferation, achieving antibody repertoire diversification and affinity maturation, and making differentiation decisions, members of this receptor family are likely to play important roles in regulating B cell maturation. If the FcRHs should prove to have differential Ig isotype Fc-binding properties, these receptors could be involved in modulating B cell differentiation so as to ensure fidelity during the maturation of somatically mutated immunoglobulin bearing cells. This prediction is based on the FcRH potential to activate stimulatory signaling pathways linking proliferation and differentiation of the B cell. Conversely, FcRH4 could have an inhibitory effect on B cell responses. *FcRH* expression and utilization during critical points in this process theoretically could also provide appropriate signals for inducing memory B cell versus plasma-cell differentiation.

The relationship between the expression of *FcγRIIB* and the different *FcRHs* will need to be explored in specific B cell subsets in order to dissect the complex signal modification of the differentiating B cells that express them. The expression of at least three of the *FcRHs*, namely *FcRH1*, *FcRH3*, and *FcRH4*, appears to be upregulated within 24 h following B cell activation via BCR ligation (Alizadeh et al. 2000). The different arrays of activating and inhibitory motifs that the FcRH possess in their cytoplasmic regions suggest the possibility that they may engage competitive signaling pathways. The FcRHs clearly contrast with their FcR relatives in their possession of multiple potential ITAM and/or ITIM that may allow the FcRH to elicit a complex array of intracellular signaling elements to finely tune the activation status of the B cells and other types of cells that express them. The production of FcRH-specific monoclonal antibodies and their use in biochemical and functional analyses should help to elucidate the afferent and efferent functions of this family of five FcRHs.

Depending on their ligand specificity these receptors could mediate responses in the innate or adaptive arms of immunity. An IgM-specific FcR has been identified on activated B cells, but this FcμR has defied molecular characterization (Sanders et al. 1987; Ohno et al. 1990). If members of the FcRH family can bind IgM or other types of natural antibodies, it is possible that the FcRH receptors can function at the interface of innate and adaptive immunity.

The fact that the multigene FcRH family resides in a chromosomal region that is commonly modified in hematopoietic malignancies is of considerable interest. Molecular analyses of the cytogenetic abnormalities in lymphoproliferative disorders, particularly those of B cell origin, have led to the identification of pathways that may enhance tumor development and progression. The majority of the translocations involving B lineage genes are viewed as ones that cause deviation of the normal B cell differentiation process (Kuppers et al. 1999). The 1q21–23 locus is one of the most frequently involved sites for chromosomal alterations, including both translocations and duplications, in hematologic malignancy (Offit et al. 1991; Cigudosa et al. 1999). This region is a recurrent hotspot for genetic changes in non-Hodgkin's lymphoma. It is particularly susceptible to changes after primary transformation. For example, secondary chromosomal 1q21–23 aberrations may frequently occur in relation to t(14;18) and t(8;14) translocations (Offit et al. 1991; Johansson et al. 1995; Whang-Peng et al. 1995).

The characterization of translocation breakpoints occurring in the chromosome 1 region in hematologic malignancies has implicated several genes in the oncogeneic process, including *AF1q*, *BCL9*, *JTB*, and *MUC1* (Tse et al. 1995; Willis et al. 1998; Hatakeyama et al. 1999; Dyomin et al. 2000; Gilles et al. 2000). Pertinent to this review is the characterization of a t(1;22)(q22;q11) translocation identified in three follicular lymphomas with t(14;18)(q32;q21) (Callanan et al. 2000). This analysis identified a breakpoint linking the *FcγRIIB* locus to the proximity of rearranged, somatically mutated IgλV region gene derived from 22q11. This resulted in *FcγRIIB* over-expression, possibly as a consequence of its juxtaposition with the 3′ Igλ enhancer element. Similar characterization of a t(1;14)(q21;q32) chromosomal translocation led to the identification of the *FcRH4/IRTA1* and *FcRH5/IRTA2* genes (Hatzivassiliou et al. 2001). Analysis of *IRTA2* expression in different types of lymphoid malignancies has indicated that this gene is frequently deregulated in follicular lymphoma and multiple myeloma cell lines with 1q21 abnormalities. A 2- to 50-fold over-expression of *IRTA2* was observed in 10 of 12 follicular lymphoma cell lines with 1q21 abnormalities, compared with only 2 of 12 follicular lymphoma cell lines without 1q21 abnormalities (Hatzivassiliou et al. 2000). *IRTA2* over-expression was found to correlate with chromosomal duplication rather than with translocation, an observation that suggests the regulation of *IRTA2* expression is very sensitive to amplifications in this region.

Expressed sequence tags (est) for four of the five FcRH can be identified in the "Lymphochip" database (Alizadeh et al. 2000). According to the "Lymphochip" microarray database, which contains multiple samples of different lymphoid malignancies, the FcRHs 1–4 are differentially ex-

Table 2. Elevated FcRH expression in B cell malignancies. FcRH expression was determined by BLAST search of the "Lymphochip" est database, identification of representative ests, and comparison of relative expression levels (ALIZADEH et al. 2000). The total number of samples for each malignancy is indicated in parentheses and the numbers of samples with elevated expression are indicated for the different FcRHs

	FcRH1	FcRH2	FcRH3	FcRH4	FcRH5
DLBCL (46)	4	12	8	8	nd
B-FL (9)	9	8	3	2	nd
B-CLL (11)	10	10	7	0	nd

B-CLL, B cell chronic lymphocytic lymphoma; B-FL, B cell follicular lymphoma; DLBCL, diffuse large B cell lymphoma; nd, no data.

pressed in chronic lymphocytic leukemias (B-CLL), follicular lymphomas (B-FL), and some of the diffuse large cell lymphomas of B lineage (DLBCL) (Table 2). Specifically, FcRH1, 2, and 3 are expressed at relatively high levels in most of the B-CLL samples represented on the array, whereas FcRH4 is not abundantly expressed in B-CLL, FcRHs 1 and 2 are over-expressed more frequently than FcRH3 and 4 in B-FL. In DLBCL, FcRH2 is over-expressed in 26% of the samples; FcRH3 and 4 are over-expressed in 17% versus 9% for FcRH1. These preliminary data on FcRH expression levels thus suggest their participation in the pathogenesis of these common lymphoproliferative disorders. Further analysis of the FcRH/IRTA family is thus likely to contribute to a better understanding of the tumors that are frequently associated with mutation in this chromosome 1 region.

The breakpoint genes so far characterized at the 1q21–23 region appear to play heterogeneous roles in B cell lymphoproliferative disorders. However, the implied functional activity and preferential B cell expression of the FcRH/IRTA receptors may serve as a gene target that is pertinent to B lymphoid malignancies. Ultimately, the *FcRH/IRTA* family may contribute to an understanding of the tumor progression that accompanies aberrations in this chromosome 1 region.

10 Concluding Remarks

The Fc receptor-related Ig-like family of molecules has attracted increasing attention in recent years. The escalating numbers of newly identified members of this family have enabled greater understanding of their phylogeny, extracellular diversity, as well as their common tyrosine-based signaling properties. Many of the functional properties of these receptors have been elucidated on the basis of molecular signaling potential that was defined previously for the classic Fc receptors that reside on chromosome 1. While many of the recently characterized pairs of Ig-like receptors have common extracellular structures, few if any have been identified as Fc receptors in the classic sense of binding immunoglobulin. Characterization of the Fc receptor homologs presents an entirely new facet of the Fc receptor family. Identification of the Fc receptor homologs on the basis of a consensus motif that is conserved in the Fc ligand-binding portion of these molecules may indicate that a relatively large number of previously uncharacterized FcR exists. These newly identified FcRH family members are related to the neighboring Fc receptors in their genomic structure, location, individual Ig-like domains, and potential ligand-binding regions. In addition, the diversity of individual tyrosine-based motifs in each of these molecules predicts an interesting array of properties for these receptors. The preferential expression of the *FcRH* in B lineage cells is thus likely to open a new chapter in B cell biology.

Acknowledgements. We thank Dr. Peter D. Burrows for critical comments on the manuscript, Marsha Flurry, Dottie Lang, and E. Ann Brookshire for help in manuscript preparation. This work has been supported in part by NIH grants AI39816 and AI47127. M.D.C. is a Howard Hughes Medical Institute Investigator. R.S.D. was supported by the Walter B. Frommeyer, Jr. Fellowship in Investigative Medicine and NIH grant DK07488.

References

Alizadeh AA, Eisen MB, Davis RE, Ma C, Lossos IS, Rosenwald A, Boldrick JC, Sabet H, Tran T, Yu X, Powell JI, Yang L, Marti GE, Moore T, Hudson J Jr, Lu L, Lewis DB, Tibshirani R, Sherlock G, Chan WC, Greiner TC, Weisenburger DD, Armitage JO, Warnke R, Staudt LM (2000) Distinct types of diffuse large B-cell lymphoma identified by gene expression profiling. Nature 403:503–511

Alley TL, Cooper MD, Chen M, Kubagawa H (1998) Genomic structure of PIR-B, the inhibitory member of the paired immunoglobulin-like receptor genes in mice. Tissue Antigens 51:224–3144

Altschul SF, Gish W, Miller W, Meyers EW, Lipman DJ (1990) Basic local alignment search tool. J Mol Biol 215:403–410

Amigorena S, Bonnerot C, Choquet D, Fridman WH, Teillaud JL (1989) FcγRII expression in resting and activated B lymphocytes. Eur J Immunol 19:1379–1385

Barten R, Torkar M, Haude A, Trowsdale J, Wilson MJ (2001) Divergent and convergent of NK-cell receptors. Trends Immunol 22:52–57

Bolland S, Ravetch JV (2000) Spontaneous autoimmune disease in FcγRIIB-deficient mice results from strain-specific epistasis. Immunity 13:277–285

Borges L, Hsu M-L, Fanger N, Kubin M, Cosman D (1997) A family of human lymphoid and myeloid Ig-like receptors, some of which bind to MHC class I molecules. J Immunol 159:5192–5196

Callanan MB, Le Baccon P, Mossuz P, Duley S, Bastard C, Hamoudi R, Dyer MJ, Klobeck G, Rimokh R, Sotto JJ, Leroux D (2000) The IgG Fc receptor, FcγRIIB, is a target for deregulation by chromosomal translocation in malignant lymphoma. Proc Natl Acad Sci USA 97:309–314

Chapman TL, Heikeman AP, Bjorkman PJ (1999) The inhibitory receptor LIR-1 uses a common binding interaction to recognize class I MHC molecules and the viral homolog UL18. Immunity 11:603–613

Chapman TL, Heikema AP, West AP Jr, Bjorkman PJ (2000) Crystal structure and ligand binding properties of the D1D2 region of the inhibitory receptor LIR-1 (ILT2). Immunity 13:727–736

Cigudosa JC, Parsa NZ, Louie DC, Filippa DA, Jhanwar SC, Johansson B, Mitelman F, Chaganti RS (1999) Cytogenetic analysis of 363 consecutively ascertained diffuse large B-cell lymphomas. Genes Chromosom Cancer 25:123–133

Clynes R, Dumitru C, Ravetch JV (1998) Uncoupling of immune complex formation and kidney damage in autoimmune glomerulonephritis. Science 279:1052–1054

Clynes RA, Towers TL, Presta LG, Ravetch JV (2000) Inhibitory Fc receptors modulate in vivo cytoxicity against tumor targets. Nature Med 6:443–446

Colonna M, Navarro F, Bellon T, Llano M, Garcia P, Samaridis J, Angman L, Cella M, Lopez-Botet M (1997) A common inhibitory receptor for major histocompatibility complex class I molecules on human lymphoid and myelomonocytic cells. J Exp Med 186:1809–1818

Cosman D, Fanger N, Borges L, Kubin M, Chin W, Peterson L, Hsu ML (1997) A novel immunoglobulin superfamily receptor for cellular and viral MHC class I molecules. Immunity 7:273–282

Daeron M (1997) Fc receptor biology. Annu Rev Immunol 15:203–234

Daeron M, Vivier E (1999) Biology of immunoreceptor tyrosine-based inhibition motif-bearing molecules. Current Topics in Microbiol Immunol 244:1–12

Daeron M, Latour S, Malbec O, Espinosa E, Pina P, Pasmans S, Fridman WH (1995) The same tyrosine-based inhibition motif, in the intracytoplasmic domain of FcγRIIB, regulates negatively BCR-, TCR-, and FcR-dependent cell activation. Immunity 3:635–646

D'Ambrosio D, Hippen KL, Minskoff SA, Mellman I, Pani G, Siminovitch KA, Cambier JC (1995) Recruitment and activation of PTP1 C in negative regulation of antigen receptor signaling by Fc gamma RIIB1. Science 268(5208):293–297

Davis RS, Wang YH, Kubagawa H, Cooper MD (2001) Identification of a family of Fc receptor homologs with preferential B cell expression. Proc Natl Acad Sci USA 98:9772–9777

Dennis G Jr, Kubagawa H, Cooper MD (2000) Paired Ig-like receptor homologs in birds and mammals share a common ancestor with mammalian Fc receptor. Proc Natl Acad Sci USA 97:13245–13250

Dyall R, Vasovic LV, Clynes RA, Nikolic-Zugic J (1999) Cellular requirements for the monoclonal antibody-mediated eradication of an established solid tumor. Eur J Immunol 29:30–37

Dyomin VG, Palanisamy N, Lloyd KO, Dyomina K, Jhanwar SC, Houldsworth J, Chaganti RS (2000) MUC1 is activated in a B-cell lymphoma by the t(1;14)(q21;q32) translocation and is rearranged and amplified in B-cell lymphoma subsets. Blood 95:2666–2671

Ernst LK, Duchemin AM, Anderson CL (1993) Association of the high-affinity receptor for IgG (Fc gamma RI) with the gamma subunit of the IgE receptor. Proc Natl Acad Sci USA 90:6023–6027

Fan QR, Mosyak L, Winter CC, Wagtmann N, Long EO, Wiley DC (1997) Structure of the inhibitory receptor for human natural killer cells resembles haematopoietic receptors. Nature 389:96–100

Fanger NA, Cosman D, Peterson L, Braddy SC, Maliszewski CR, Borges L (1998) The MHC class I binding proteins LIR-1 and LIR-2 inhibit Fc receptor-mediated signaling in monocytes. Eur J Immunol 28:3423–3434

Garman SC, Kinet J-P, Jardetzky TS (1998) Crystal structure of the human high-affinity IgE receptor. Cell 95:951–961

Garman SC, Wurzburg BA, Tarchevskaya SS, Kinet JP, Jardetzky TS (2000) Structure of the Fc fragment of human IgE bound to its high-affinity receptor FcεRIα. Nature 406:259–266

Gergely J, Pecht I, Sarmay G (1999) Immunoreceptor tyrosine-based inhibition motif-bearing receptors regulate the immunoreceptor tyrosine-based activation motif-induced activation of immune competent cells. Immunol Lett 68: 3–15

Gilles F, Goy A, Remache Y, Shue P, Zelenetz AD (2000) MUC1 dysregulation as the consequence of a t(1;14)(q21;q32) translocation in an extranodal lymphoma. Blood 95:2930–2936

Guex N, Peitsch MC (1997) SWISS-MODEL and the Swiss-Pdb viewer: an environment for comparative protein modeling. Electrophoresis 18:2714–2723

Hatakeyama S, Osawa M, Omine M, Ishikawa F (1999) JTB: a novel membrane protein gene at 1q21 rearranged in a jumping translocation. Oncogene 18:2085–2090

Hatzivassiliou G, Miller I, Takizawa J, Palanisamy N, Rao PH, Iida S, Tagawa S, Taniwaki M, Russo J, Neri A, Catoretti G, Clynes R, Mendlsohn C, Chaganti RSK, Dalla-Favera, R (2001) IRTA1 and IRTA2, novel immunoglobulin superfamily receptors expressed in B cells and involved in chromosome 1q21 abnormalities in B cell malignancy. Immunity 14:277–289

Hayami K, Fukuta D, Nishikawa Y, Yamashita Y, Inu M, Ohyama Y, Hikida M, Ohmori H, Takai T (1997) Molecular cloning of a novel murine cell surface glycoprotein homologous to killer cell inhibitory receptors. J Biol Chem 272:7320–7327

Henry C, Jerne NK (1968) Competition of 19 S and 7 S antigen receptors in the regulation of the primary immune response. J Exp Med 128:133–152

Heyman B (1999) Antibody feedback suppression: towards a unifying concept? Immunol Lett 68:41–45

Heyman B (2000) Regulation of antibody responses via antibodies, complement, and Fc receptors. Annu Rev Immunol 18:709–737

Hibbs ML, Selvaraj P, Carpen O, Springer TA, Kuster H, Jouvin MH, Kinet JP (1989) Mechanisms for regulating expression of membrane isoforms of Fc gamma RIII (CD16). Science 246:1608–1611

Ho LH, Uehara T, Chen CC, Kubagawa H, Cooper MD (1999) Constitutive tyrosine phosphorylation of the inhibitory paired Ig-like receptor PIR-B. Proc Natl Acad Sci USA 96:15086–15090

Hulett MD, Witort E, Brinkworth RI, McKenzie IFC, Hogarth PM (1995) Multiple regions of human FcγRII (CD32) contribute to the binding of IgG. J Biol Chem 270:21188–21194

Johansson B, Mertens F, Mitelman F (1995) Cytogenetic evolution patterns in non-Hodgkin's lymphoma. Blood 86:3905–3914

Kasahara M (1999) The chromosomal duplication model of the major histocompatibility complex. Immunol Rev 167:17–32

Kleinau S, Martinsson P, Heyman B (2000) Induction and suppression of collagen-induced arthritis is dependent on distinct FcγR. J Exp Med 191:1611–1616

Krajci P, Grzeschik KH, Geurts van Kessel AH, Olaisen B, Brandtzaeg P (1991) The human transmembrane secretory component (poly-Ig receptor): molecular cloning, restriction fragment length polymorphism and chromosomal sublocalization. Hum Genet 87: 642–648.

Kremer EJ, Kalatzis V, Baker E, Callen DF, Sutherland GR, Maliszewski CR (1992) The gene for the human IgA Fc receptor maps to 19q13.4. Hum Genet 89:107–108

Kubagawa H, Burrows P, Cooper MD (1997) A novel pair of immunoglobulin-like receptors expressed by B cells and myeloid cells. Proc Natl Acad Sci USA 94:5261–5266

Kulczycki A Jr, Webber J, Soares HA, Onken MD, Thompson JA, Chaplin DD, Loh DY (1990) Genomic organization of the mouse Fcγ receptor genes. Proc Natl Acad Sci USA 87:2856–2860

Kuppers R, Klein U, Hansmann ML, Rajewsky K (1999) Cellular origin of human B-cell lymphomas. N Engl J Med 341:1520–1529

Lanier LL (1998a) NK cell receptors. Annu Rev Immunol 16:359–393

Lanier LL (1998b) Follow the leader: NK cell receptors for classical and non-classical MHC class I. Cell 92:705–707

Long EO (1999) Regulation of immune responses through inhibitory receptors. Annu Rev Immunol 17:875–904

Maxwell KF, Powell MS, Hulett MD, Barton PA, McKenzie IF, Garrett TP, Hogarth PM (1999) Crystal structure of the human leukocyte Fc receptor, FcγRIIA. Nat Struct Biol 6:437–442

Moretta A, Bottino C, Vitale M, Pende D, Biassoni R, Mingari MC, Moretta L (1996) Receptors for HLA class-I molecules in human natural killer cells. Annu Rev Immunol 14:619–648

Morton HC, Zandbergen van G, Kooten van C, Howard CJ, Winkel van de JG, Brandtzaeg P (1999) Immunoglobulin-binding sites of human FcαRI (CD89) and bovine Fcγ2R are located in their membrane-distal extracellular domains. J Exp Med 189:1715–1722

Morton HC, Van den Herik-Oudijk IE, Vossebeld P, Snijders A, Verhoeven AJ, Capel PJA, Winkel van de JGJ (1995) Functional association between the human myeloid immunoglobulin A Fc receptor (CD89) and FcRγ chain. Molecular basis for CD89/FcRγ chain association. J Biol Chem 270:29781–29787

Nakamura A, Yuasa T, Ujike A, Ono M, Nukiwa T, Ravetch JV, Takai T (2000) FcγRIIB-deficient mice develop Goodpasture's syndrome upon immunization with type IV collagen: a novel murine model for autoimmune glomerular basement membrane disease. J Exp Med 191:899–906

Nicholas KB, Nicholas HB Jr, Deerfield DW II (1997) EMBnet News 4:14

Oakley RJ, Howard TA, Hogarth PM, Tani K, Seldin MF (1992) Chromosomal mapping of the high affinity Fcγ receptor gene. Immunogenetics 35:279–282

Offit K, Wong G, Filippa DA, Tao Y, Chaganti RS (1991) Cytogenetic analysis of 434 consecutively ascertained specimens of non-Hodgkin's lymphoma: clinical correlations. Blood 77:1508–1515

Ohno T, Kubagawa H, Sanders SK, Cooper MD (1990) Biochemical nature of an Fcμ receptor on human B-lineage cells. J Exp Med 172:1165–1175

Ono M, Bolland S, Tempst P, Ravetch JV (1996) Role of the inositol phosphatase SHIP in negative regulation of the immune system by the receptor Fc(gamma)RIIB Nature 383:263–266

Pang J, Taylor GR, Munroe DG, Ishaque A, Fung-Leung WP, Lau CY, Liu FT, Zhou L (1994) Characterization of the gene for the high affinity IgE receptor (FcεRI) α-chain. J Immunol 151:6166–6174

Pascual V, Liu Y-J, Magalski A, de Bouteiller O, Bancereau J, Capra JD (1994) Analysis of somatic mutation in five B cell subsets of human tonsil. J Exp Med 180:329–339

Perez-Montfort R, Kinet JP, Metzger H (1983) A previously unrecognized subunit of the receptor for immunoglobulin E. Biochemistry 22:5722–5728

Pfefferkorn LC, Yeaman GR (1994) Association of IgA-Fc receptors (FcαR) with FcεRIγ2 subunits in U937 cells. Aggregation induces the tyrosine phosphorylation of γ2. J Immunol 153:3228–3236

Pleass RJ, Dunlop JI, Anderson CM, Woof JM (1999) Identification of residues in the CH2/CH3 domain interface of IgA essential for interaction with the human Fcα receptor (FcαR) CD89. J Biol Chem 274:23508–23514

Qui WQ, de Bruin D, Brownstein BH, Pearse R, Ravetch JV (1990) Organization of the human and mouse low-affinity FcγR genes: amplification and recombination Science 248:732–735

Ravetch JV, Kinet J-P (1991) Fc receptors. Annu Rev Immunol 9:457–492

Ravetch JV, Clynes RA (1998) Divergent roles for Fc receptors and complement in vivo. Annu Rev Immunol 16:421–432

Ravetch JV, Lanier LL (2000) Immune inhibitory receptors. Science 290:84–89

Ravetch JV, Bolland S (2001) IgG Fc receptors. Annu Rev Immunol 19:275–290

Reth M (1992) Antigen receptors on B lymphocytes. Annu Rev Immunol 10:97–121

Samaridis J, Colonna M (1997) Cloning of novel immunoglobulin superfamily receptors expressed on human myeloid and lymphoid cells: Structural evidence for new stimulatory and inhibitory pathways. Eur J Immunol 27:660–665

Sanders SK, Kubagawa H, Suzuki T, Butler JL, Cooper MD (1987) IgM binding protein expressed by activated B cells. J Immunol 139:188–193

Scholl PR, Geha RS (1993) Physical association between the high-affinity IgG receptor (Fc gamma RI) and the gamma subunit of the high-affinity IgE receptor (Fc epsilon RI gamma). Proc Natl Acad Sci USA 90:8847–8850

Selvakumar A, Steffens U, Palansami N, Chaganti RS, Dupont B (1997) Genomic organization and allelic polymorphism of the human killer cell inhibitory receptor gene KIR103. Tissue Antigens 49:564–573

Shibuya A, Sakamoto N, Shimizu Y, Shibuya K, Osawa M, Hiroyama T, Eyre HJ, Sutherland GR, Endo Y, Fujita T, et al (2000) Fc alpha/mu mediates endocytosis of IgM-coated microbes. Nature Immunol 1:441–446

Sondermann P, Huber R, Jacob U (1999) Crystal structure of the soluble form of the human FcγRIIB: a new member of the immunoglobulin superfamily at 1.7 A resolution. Embo J 18:1095–1103

Sondermann P, Kaiser J, Jacob U (2001) Molecular Basis for Immune Complex Recognition: a comparison of Fc-receptor Structures. J Mol Biol 309:737–749

Sondermann P, Huber R, Oosthuizen V, Jacob U (2000) The 3.2-A crystal structure of the human IgG1 Fc fragment-FcγRIII complex. Nature 406:267–273

Sylvestre DL, Ravetch JV (1996) A dominant role for mast cell Fc receptor in the Arthus reaction. Immunity 5:387–390

Tamm A, Schmidt RE (1997) IgG binding sites on human Fcγ receptors. Int Rev Immunol 16:57–85

Tepler I, Morton CC, Shimizu A, Holcombe RF, Eddy R, Shows TB, Leder P (1989) The gene for the human mast cell high-affinity IgE receptor alpha chain: chromosomal locaticalization to 1q21- of 23 and RFLP analysis. Am J Hum Genet 45:761–765

Thompson JD, Gibson TJ, Plewniak F, Jeanmougin F, Higgins DG (1997) The CLUSTAL-X windows interface: flexible strategies for multiple sequence alignment aided by quality analysis tools. Nucleic Acids Res 25:4876–4882

Torkar M, Norgate Z, Colonna M, Trowsdale J, Wilson MJ (1998) Isotypic variation of the novel immunoglobulin-like transcript/killer cell inhibitory receptor loci in the leukocyte receptor complex. Eur J Immunol 28: 3959–3967

Tse W, Zhu W, Chen HS, Cohen A (1995) A novel gene, AF1q, fused to MLL in t(1;11) (q21;q23), is specifically expressed in leukemic and immature hematopoietic cells. Blood 85:650–656

Unkeless JC, Jin J (1997) Inhibitory receptors, ITIM sequences and phosphatases. Curr Opin Immunol 9:338–343

Van de Peer Y, De Wachter R (1993) TREECON: a software package for the construction and drawing of evolutionary trees. Comp Appl Biosci 9:177–182

Vely F, Vivier E (1997) Conservation of structural features reveals the existence of a large family of inhibitory cell surface receptors and noninhibitory/activatory counterparts. J Immunol 159:2075–2077

Wagtmann N, Rojo S, Eichler E, Mohrenweiser H, Long EO (1997) A new human gene complex encoding the killer cell inhibitory receptors and related monocyte/macrophage receptors. Curr Biol 7:615–618

Wende H, Colonna M, Ziegler A, Volz A (1999) organization of the leukocyte receptor cluster (LRC) on human chromosome 19q13.4. Mamm Genome 10:154–160

Whang-Peng J, Knutsen T, Jaffe ES, Steinberg SM, Raffeld M, Zhao WP, Duffey P, Condron K, Yano T, Longo DL (1995) Sequential analysis of 43 patients with non-Hodgkin's lymphoma: clinical correlations with cytogenetic, histologic, immunophenotyping, and molecular studies. Blood 85:203–216

Williams AF, Barclay AN (1988) The immunoglobulin superfamily–domains for cell surface recognition. Annu Rev Immunol 6:381–405

Willis TG, Zalcberg IR, Coignet LJ, Wlodarska I, Stul M, Jadayel DM, Bastard C, Treleaven JG, Catovsky D, Silva ML, Dyer MJ (1998) Molecular cloning of translocation t(1;14)(q21;q32) defines a novel gene (BCL9) at chromosome 1q21. Blood 91:1873–1881

Wilson MJ, Torkar M, Trowsdale J (1997) Genomic organization of a human killer cell inhibitory receptor gene Tissue Antigens 49:574–579

Wilson MJ, Torkar M, Haude A, Milne S, Jones T, Sheer D, Beck S, Trowsdale J (2000) Plasticity in the organization and sequences of human KIR/ILT gene families. Proc Natl Acad Sci USA 97:4778–4783

Winkel van de JG, Capel PJ (1993) Human IgG Fc receptor heterogeneity: molecular aspects and clinical implications. Immunol Today 14:215–221

Wit de TPM, Morton HC, Capel PJA, Winkel van de JGJ (1995) Structure of the gene for the human myeloid IgA Fc receptor (CD89). J Immunol 155:1203–1209

Xu M, Zhao R, Zhao ZJ (2001) Molecular cloning and characterization of SPAP1, an inhibitory receptor. Biochem Biophys Res Commun 280:768–775

Ye ZS, Kinet JP, Paul WE (1992) Structure of the gene for the α-chain of the mouse high affinity receptor for IgE (FcεRI). J Immunol 149:897–900

Yuasa T, Kubo S, Yoshino T, Ujike A, Matsumura K, Ono M, Ravetch JV, Takai T (1999) Deletion of FcγIIB renders H-2(b) mice susceptible to collagen-induced arthritis. J Exp Med 189:187–194

Zhang G, Young JR, Tregaskes CA, Sopp P, Howard CJ (1995) Identification of a novel class of mammalian FcγR. J Immunol 155:1534–1541

Subject Index

Printing (Computer to Film): Saladruck Berlin
Binding: Stürtz AG, Würzburg

Current Topics in Microbiology and Immunology

Volumes published since 1989 (and still available)

Vol. 223: **Tracy, S.; Chapman, N. M.; Mahy, B. W. J. (Eds.):** The Coxsackie B Viruses. 1997. 37 figs. VIII, 336 pp. ISBN 3-540-62390-6

Vol. 224: **Potter, Michael; Melchers, Fritz (Eds.):** C-Myc in B-Cell Neoplasia. 1997. 94 figs. XII, 291 pp. ISBN 3-540-62892-4

Vol. 225: **Vogt, Peter K.; Mahan, Michael J. (Eds.):** Bacterial Infection: Close Encounters at the Host Pathogen Interface. 1998. 15 figs. IX, 169 pp. ISBN 3-540-63260-3

Vol. 226: **Koprowski, Hilary; Weiner, David B. (Eds.):** DNA Vaccination/Genetic Vaccination. 1998. 31 figs. XVIII, 198 pp. ISBN 3-540-63392-8

Vol. 227: **Vogt, Peter K.; Reed, Steven I. (Eds.):** Cyclin Dependent Kinase (CDK) Inhibitors. 1998. 15 figs. XII, 169 pp. ISBN 3-540-63429-0

Vol. 228: **Pawson, Anthony I. (Ed.):** Protein Modules in Signal Transduction. 1998. 42 figs. IX, 368 pp. ISBN 3-540-63396-0

Vol. 229: **Kelsoe, Garnett; Flajnik, Martin (Eds.):** Somatic Diversification of Immune Responses. 1998. 38 figs. IX, 221 pp. ISBN 3-540-63608-0

Vol. 230: **Kärre, Klas; Colonna, Marco (Eds.):** Specificity, Function, and Development of NK Cells. 1998. 22 figs. IX, 248 pp. ISBN 3-540-63941-1

Vol. 231: **Holzmann, Bernhard; Wagner, Hermann (Eds.):** Leukocyte Integrins in the Immune System and Malignant Disease. 1998. 40 figs. XIII, 189 pp. ISBN 3-540-63609-9

Vol. 232: **Whitton, J. Lindsay (Ed.):** Antigen Presentation. 1998. 11 figs. IX, 244 pp. ISBN 3-540-63813-X

Vol. 233/I: **Tyler, Kenneth L.; Oldstone, Michael B. A. (Eds.):** Reoviruses I. 1998. 29 figs. XVIII, 223 pp. ISBN 3-540-63946-2

Vol. 233/II: **Tyler, Kenneth L.; Oldstone, Michael B. A. (Eds.):** Reoviruses II. 1998. 45 figs. XVI, 187 pp. ISBN 3-540-63947-0

Vol. 234: **Frankel, Arthur E. (Ed.):** Clinical Applications of Immunotoxins. 1999. 16 figs. IX, 122 pp. ISBN 3-540-64097-5

Vol. 235: **Klenk, Hans-Dieter (Ed.):** Marburg and Ebola Viruses. 1999. 34 figs. XI, 225 pp. ISBN 3-540-64729-5

Vol. 236: **Kraehenbuhl, Jean-Pierre; Neutra, Marian R. (Eds.):** Defense of Mucosal Surfaces: Pathogenesis, Immunity and Vaccines. 1999. 30 figs. IX, 296 pp. ISBN 3-540-64730-9

Vol. 237: **Claesson-Welsh, Lena (Ed.):** Vascular Growth Factors and Angiogenesis. 1999. 36 figs. X, 189 pp. ISBN 3-540-64731-7

Vol. 238: **Coffman, Robert L.; Romagnani, Sergio (Eds.):** Redirection of Th1 and Th2 Responses. 1999. 6 figs. IX, 148 pp. ISBN 3-540-65048-2

Vol. 239: **Vogt, Peter K.; Jackson, Andrew O. (Eds.):** Satellites and Defective Viral RNAs. 1999. 39 figs. XVI, 179 pp. ISBN 3-540-65049-0

Vol. 240: **Hammond, John; McGarvey, Peter; Yusibov, Vidadi (Eds.):** Plant Biotechnology. 1999. 12 figs. XII, 196 pp. ISBN 3-540-65104-7

Vol. 241: **Westblom, Tore U.; Czinn, Steven J.; Nedrud, John G. (Eds.):** Gastroduodenal Disease and Helicobacter pylori. 1999. 35 figs. XI, 313 pp. ISBN 3-540-65084-9

Vol. 242: **Hagedorn, Curt H.; Rice, Charles M. (Eds.):** The Hepatitis C Viruses. 2000. 47 figs. IX, 379 pp. ISBN 3-540-65358-9

Vol. 243: **Famulok, Michael; Winnacker, Ernst-L.; Wong, Chi-Huey (Eds.):** Combinatorial Chemistry in Biology. 1999. 48 figs. IX, 189 pp. ISBN 3-540-65704-5

Vol. 244: **Daëron, Marc; Vivier, Eric (Eds.):** Immunoreceptor Tyrosine-Based Inhibition Motifs. 1999. 20 figs. VIII, 179 pp. ISBN 3-540-65789-4

Vol. 245/I: **Justement, Louis B.; Siminovitch, Katherine A. (Eds.):** Signal Transduction and the Coordination of B Lymphocyte Development and Function I. 2000. 22 figs. XVI, 274 pp. ISBN 3-540-66002-X

Vol. 245/II: **Justement, Louis B.; Siminovitch, Katherine A. (Eds.):** Signal Transduction on the Coordination of B Lymphocyte Development and Function II. 2000. 13 figs. XV, 172 pp. ISBN 3-540-66003-8

Vol. 246: **Melchers, Fritz; Potter, Michael (Eds.):** Mechanisms of B Cell Neoplasia 1998. 1999. 111 figs. XXIX, 415 pp. ISBN 3-540-65759-2

Vol. 247: **Wagner, Hermann (Ed.):** Immunobiology of Bacterial CpG-DNA. 2000. 34 figs. IX, 246 pp. ISBN 3-540-66400-9

Vol. 248: **du Pasquier, Louis; Litman, Gary W. (Eds.):** Origin and Evolution of the Vertebrate Immune System. 2000. 81 figs. IX, 324 pp. ISBN 3-540-66414-9

Vol. 249: **Jones, Peter A.; Vogt, Peter K. (Eds.):** DNA Methylation and Cancer. 2000. 16 figs. IX, 169 pp. ISBN 3-540-66608-7

Vol. 250: **Aktories, Klaus; Wilkins, Tracy, D. (Eds.):** Clostridium difficile. 2000. 20 figs. IX, 143 pp. ISBN 3-540-67291-5

Vol. 251: **Melchers, Fritz (Ed.):** Lymphoid Organogenesis. 2000. 62 figs. XII, 215 pp. ISBN 3-540-67569-8

Vol. 252: **Potter, Michael; Melchers, Fritz (Eds.):** B1 Lymphocytes in B Cell Neoplasia. 2000. XIII, 326 pp. ISBN 3-540-67567-1

Vol. 253: **Gosztonyi, Georg (Ed.):** The Mechanisms of Neuronal Damage in Virus Infections of the Nervous System. 2001. approx. XVI, 270 pp. ISBN 3-540-67617-1

Vol. 254: **Privalsky, Martin L. (Ed.):** Transcriptional Corepressors. 2001. 25 figs. XIV, 190 pp. ISBN 3-540-67569-8

Vol. 255: **Hirai, Kanji (Ed.):** Marek's Disease. 2001. 22 figs. XII, 294 pp. ISBN 3-540-67798-4

Vol. 256: **Schmaljohn, Connie S.; Nichol, Stuart T. (Eds.):** Hantaviruses . 2001, 24 figs. XI, 196 pp. ISBN 3-540-41045-7

Vol. 257: **van der Goot, Gisou (Ed.):** Pore-Forming Toxins, 2001. 19 figs. IX, 166 pp. ISBN 3-540-41386-3

Vol. 258: **Takada, Kenzo (Ed.):** Epstein-Barr Virus and Human Cancer. 2001. 38 figs. IX, 233 pp. ISBN 3-540-41506-8

Vol. 259: **Hauber, Joachim, Vogt, Peter K. (Eds.):** Nuclear Export of Viral RNAs. 2001. 19 figs. IX, 131 pp. ISBN 3-540-41278-6

Vol. 260: **Burton, Didier R. (Ed.):** Antibodies in Viral Infection. 2001. 51 figs. IX, 309 pp. ISBN 3-540-41611-0

Vol. 261: **Trono, Didier (Ed.):** Lentiviral Vectors. 2002. 32 figs. X, 258 pp. ISBN 3-540-42190-4

Vol. 262: **Oldstone, Michael B.A. (Ed.):** Arenaviruses I. 2002, 30 figs. XVIII, 197 pp. ISBN 3-540-42244-7

Vol. 263: **Oldstone, Michael B. A. (Ed.):** Arenaviruses II. 2002, 49 figs. XVIII, 268 pp. ISBN 3-540-42705-8

Vol. 264/I: **Hacker, Jörg; Kaper, James B. (Eds.):** Pathogenicity Islands and the Evolution of Microbes. 2002. 34 figs. XVIII, 232 pp. ISBN 3-540-42681-7

Vol. 264/II: **Hacker, Jörg; Kaper, James B. (Eds.):** Pathogenicity Islands and the Evolution of Microbes. 2002. 24 figs. XVIII, 228 pp. ISBN 3-540-42682-5

Vol. 265: **Dietzschold, Bernhard; Richt, Jürgen A. (Eds.):** Protective and Pathological Immune Responses in the CNS. 2002. 21 figs. X, 278 pp. ISBN 3-540-42668-X